Trust Your Gut

Trust Your Gut

Get Lasting Healing from IBS and Other Chronic Digestive Problems Without Drugs

---

Gregory A. Plotnikoff, MD, MTS, FACP
&
Mark B. Weisberg, PhD, ABPP

with Steve LeBeau

Conari Press

First published in 2013 by Conari Press, an imprint of
Red Wheel/Weiser, LLC
With offices at:
665 Third Street, Suite 400
San Francisco, CA94107
www.redwheelweiser.com

ISBN: 978-1-57324-588-3

Library of Congress Cataloging-in-Publication data available upon request

Cover design by Jim Warner
Cover photograph © *Zen Shui/SuperStock*
Interior by Maureen Forys, Happenstance Type-O-Rama
Typeset in New Century Schoolbook and Optima

Printed in the United States of America
VG

10 9 8 7 6 5 4 3 2 1

The paper used in this publication meets the minimum requirements of the
American National Standard for Information Sciences—Permanence of Paper
for Printed Library Materials Z39.48-1992 (R1997).

CONTENTS

We dedicate this book to our many teachers: patients, professors, mentors, colleagues, friends, and family. We are deeply grateful for all that you have taught us. You have blessed us. We hope this book honors you.

INTRODUCTION

We created this book to empower you to take control of your health.

If you or someone you love is plagued by chronic digestive distress, you know what it's like to be held captive by your gut. You feel increasingly frustrated and don't know what to do next. You've probably seen several competent physicians who have devoted their best efforts to thoroughly diagnose and treat your condition. You may have tried many different medications to treat the symptoms, only to find that they brought temporary relief at best. You have lived in fear of your unpredictable "problem" that forces you to find excuses for canceling events. We wrote this book to let you know there is a way out of this cycle of suffering. We have combined ancient wisdom about the body and mind with the newest findings in medicine, psychology, and neuroscience to create a holistic program that lets you take back your life. All you need to add to the equation is you.

Here is a story that may be familiar to you. We have heard variations of it from hundreds of patients. It is the story of "Maria," who is a composite of many people we have treated.

Maria's Helplessness

Maria, a married mother of three teenagers, worked full time in a busy office downtown. One day she began to feel pains in her abdomen, followed by a severe bout of diarrhea. She figured it would go away, like it always had before.

"I was too busy to care about me," she said.

But it didn't go away. It got worse. The diarrhea started alternating with periods of constipation. She couldn't predict what she would experience next. She began missing meetings at work because she was in the restroom. Some days she would call in sick "just in case."

When she realized she was distancing herself from her family and friends, she went to see her primary care physician. She examined Maria and ran all the appropriate diagnostic tests but couldn't find anything physically wrong. She also assessed her for depression, anxiety, post-traumatic stress disorder, and excessive stress. Maria's doctor recommended fiber and gave her a referral to see a gastroenterologist (GI specialist). The GI specialist gave Maria a colonoscopy and several other tests to rule out serious physical pathology. "The results are all negative," said the specialist. "This means that your symptoms are not due to something serious like cancer or inflammatory bowel disease. Your symptoms are due to irritable bowel syndrome, or IBS."

She then discussed lifestyle changes and stress reduction, and reviewed the range of prescription drugs that might be helpful. She also suggested that Maria might benefit by seeing a psychologist to address some of the stresses affecting her symptoms. Maria was concerned about the drug recommendations because of her previous sensitivity to medications. And the suggestion to see a psychologist pressed the wrong button in her. Now Maria was really scared. "They think I'm crazy," she said to herself. "They think it's all in my head."

The psychologist could not solve her problem either. "It is true that severe depression or acute anxiety can cause intestinal problems," he told Maria, "but I don't think you have psychological issues significant enough to warrant treatment. A psychiatrist might prescribe something to help you relax, but otherwise I don't think my services would be helpful for you at this time."

Maria was back where she started, only worse. She made the rounds of seeing different doctors and getting prescriptions that treated her symptoms, but she felt no hope for successful treatment. "I've been scoped from both ends and told that not much can be done for me," said Maria. "Yet I still have pain and gastric distress. What now?"

What Maria didn't realize was that she wasn't the only one who felt frustrated and helpless. The health professionals who see patients like Maria often feel frustrated by their limited impact on problems like hers. This can be true even for GI specialists, the physicians with the most training in gastrointestinal diseases. They are the ones we go to for evaluation for the most serious or life-threatening diseases, such as gastrointestinal cancer and inflammatory bowel disease. They have the technologies for diagnosis of organic or structural diseases. But even they get frustrated by the severe suffering experienced by those who have unexplained symptoms, what is termed functional bowel disease. In spite of the diligent and caring work of GI specialists, Maria and millions of others can't find relief for their chronic digestive distress. Can't anything be done for them?

That's where we enter the picture. Greg Plotnikoff is an MD and a leader in the field of integrative medicine who spent six years in Japan studying traditional medicine. Mark Weisberg is a PhD psychologist who specializes in the treatment of chronic pain and the emerging field of clinical health psychology. Our novel approach extends the range of standard practice in both medicine and clinical psychology. We share a holistic vision of how the body and mind work together, a perspective that allows us to see new ways to solve old problems. As recently as ten years ago, our holistic approach would have been marginalized by the medical profession. These days we are the go-to doctors that other doctors refer their patients to when they run out of answers. We wrote *Trust Your Gut* to share our answers with you, so you can help yourself. This book will help you free yourself from your chronic misery.

A Revolution in the Treatment of Gut Distress: The CORE Program

We approach the treatment of gut issues from the premise that the mind and body are all part of an integrated system. We know both from our own clinical experience and from research data that the mind-body relationship is interactive in both directions, and we must always look at health from a 360-degree perspective. Otherwise, we miss some of the most important cues and clues to our wellness. In fact, both the latest neuroscience of the gut and the ancient wisdom of Asian medicine agree that the gut is the focal point of human energy and the seat of the emotions. Indeed, scientists are increasingly referring to the gut as the second brain. Although your gut appears to be the cause of all your problems, it is actually the center of hope for relief from your symptoms.

Western philosophy and science—starting from the days of Plato and Aristotle—have seen the mind and rational thought as part of some higher reality, whereas the body and emotions are of lesser importance. Classical philosophers taught that reason must control the emotions and that the mind must rise above bodily concerns. Centuries later, French philosopher René Descartes formalized the split by declaring the mind and body to be two metaphysically different kinds of realities. This led to centuries of scientific exploration of the body with little regard to the mind. Although few scientists believe in such dualities anymore, the study of the mind still lags far behind the study of the body. That's because it's much easier to study the body. You can see it, measure it, touch it, and x-ray it. You can do none of those things with a mind.

The problem is that our mind is subjective, but science is only looking for objective truth that can be measured. That's how behaviorism, the theory that all behavior is based on conditioning, became the dominant movement in psychology in the 20th century—it removed the mind as an object of study and focused only on behavior. This started to change by the 1970s when pioneering scientists integrated the study of psychology, neurology, and immunology—termed psychoneuroimmunology—to create the new science of mind-body interactions. It is still a very new and developing science, and we are among the first wave of health professionals to apply this new knowledge to solve chronic gastric distress. The results we've had are astounding.

Instead of talking about the body and mind as two separate entities, we talk about the body/mind. Each person is a unified system and should be approached as such. This shift in perspective was possible due to the technical advances in imaging that allow scientists to measure the brain's functional activity in living people. The most surprising insight is that our brain does not distinguish between what is physical and what is psychological. It creates the same neurohormonal responses either way. This new perspective allows a completely different way of looking at the problem of gastric distress. More important, it makes it possible to find new solutions.

The Western approach to disease and illness uses a lot of violent metaphors that suggest health care is a huge battle. Doctors fight disease, they wage war on cancer, and patients struggle valiantly to conquer the disease. The medical arsenal includes lasers, radiation, chemicals, and pills. One of the primary goals is to kill pain.

Surprisingly, in other cultures, such as that of Japan, not a single word associated with care is related with violence. In Japanese, the key actions that health professionals take are expressed with words conveying comfort, harmony, and balance. We have pain *killers,* and they have pain *calmers* or *suppressors.*

The Western "us versus them" strategy works well for a lot of illnesses, such as when you need an antibiotic to kill bacteria or chemotherapy to kill cancer cells. But this approach falls short for many gut sufferers. You probably believe that your gut is a problem to be attacked, because that is how you have been taught to think about illness. But as you can see from the title of our book, we have a totally different approach: we don't want you to *fight* your gut; we want you to *trust* it.

The main theme of this book—and the key to solving your gut distress—is that your gut is not your enemy; on the contrary, it is the center of your body/mind system. It is your core. Your chronic gut problems are signs that your system is out of balance. To restore that balance and become centered, you must learn to listen to what your gut is telling you. Just as heat sensations tell you to take your hand off the stove, and the bad smell of spoiled milk tells you not to drink it, the various symptoms of gut distress are messages that need to be deciphered and acted upon. Instead of killing the pain with a pill, we want you to observe the pain and try to understand what it is telling you.

Think about it. If you killed the pain in your hand when it was on the stove, it would become severely burned. If you killed the ability to smell, you could get sick or even worse from eating spoiled food. So why try to kill the pain and discomfort in your gut? It doesn't make the problem go away. It only allows you to forget about it.

Your gut is not your enemy; on the contrary, it is the center of your body/mind system. It is your core. Your chronic gut problems are signs that your system is out of balance.

Your body is trying to tell you something, and the best thing to do is to listen to it.

Think of all the pills and remedies you have taken over the years. They haven't brought lasting relief. They may have even created other health problems. It's time to quit fighting and start listening. Your gut is not your enemy. Your gut is part of you. You don't need *pills,* you need *skills* to help you observe and respond effectively.

It may feel like your gut is holding you hostage and is trying to sabotage your life. Yet no matter how hard you fight, you don't seem to win. It's like those old Chinese handcuffs you might have played with as a child. The woven bamboo tube slips over both of your index fingers, and the harder you try to pull them out, the tighter the cuff gets. The solution is to quit struggling and relax. Then it slips off very easily. That's very similar to our strategy. We propose to teach you a nonviolent approach to centering your body/mind system.

Trust Your Gut is the first book to address intestinal distress from this perspective—a viewpoint that sees the gut as a vital messenger to *heed and trust,* rather than as an enemy to fear. Your symptoms are messages from the body to rely on, rather than to ignore or medicate. We focus on those functional relationships between the brain and the gut—including the neural and hormonal interactions—as well as the interactions with each patient's inner and outer environments. Because we know that everything is connected, we must pay attention to all aspects of a person's life to pinpoint possible areas where the system isn't functioning properly.

Your symptoms are messages from the body to rely on, rather than to ignore or medicate.

Two Powerful New Tools for Gut Healing

This book introduces two powerful new approaches to reducing chronic gut distress: Neurohormonal Retraining and Ecological Rebalancing. These techniques calm the hypersensitivity of the digestive tract and the nervous system, a key factor that prolongs gut distress into a chronic ailment. Pills may temporarily block the symptoms, but they mask hypersensitivity instead of addressing it. Ecological Rebalancing restores the body's natural harmony by subduing external causes of stress and their internal consequences. Neurohormonal Retraining uses your power of concentration to short-circuit the hypersensitive feedback loop between the digestive system and the nervous system, which reduces both pain and gut dysfunction.

These techniques are effective because they are based on the holistic understanding that everything is connected and your gut goes haywire when these connections are imbalanced. These two approaches are the foundation of our systematic approach to gut distress that we call the CORE program.

The main components of our program are summed up in our acronym: CORE.

Center

Observe

Restore

Ensure

The CORE self-help program provides a new set of tools beyond the pharmaceutical shelves, beyond tests and conventional wisdom—tools under your control. True primary care is not what a physician does for you but rather what you do for yourself, and this book will enable you to do that. CORE will teach you new ways to deal with your gut and to break the vicious cycle of symptoms, fear, and pain.

Are You Ready to Use This Book?

If you have not yet seen your primary care physician and/or a GI specialist to have your intestinal pains and discomforts evaluated, you are not ready for this book. Some potential problems are extremely serious and even life threatening, and the health care system is excellent at recognizing and treating them. Your doctors or specialists can run tests to find out what is wrong, and then they can work with you on the problem. That is why you should first go to your physician and be thoroughly examined. (Please see our website *www.trustyourgutbook.com* for a list of conditions that require other treatments before beginning the CORE program.)

After you have been comprehensively tested and all organic, structural physiological problems have been ruled out, *then* you are ready to use this book. When your doctors tell you that they find nothing wrong on your diagnostic tests—yet your symptoms, pain, and discomfort continue—that's when we can help you. Clearly, you know something is wrong after you've been suffering for five, ten, or twenty years. The good news is that these chronic maladies are not life threatening. The great news is that there is something you can do to diminish your distress or make it go away altogether. You are ready to start trusting your gut!

This book is for you if you have been diagnosed with:

- IBS—Irritable bowel syndrome
- GERD—Gastroesophageal reflux disease (the type caused by functional disorder)
- Dysfunctional bowel

Or have any of these chronic symptoms:

- Bad breath despite good dental hygiene
- Burping, belching, hiccuping
- Esophagitis, reflux, heartburn
- Stomach upset, gastritis
- Nausea, with or without food
- Food intolerances
- Appetite concerns (too much, too little, hungry but can't eat, full too quickly)
- Bloating
- Abdominal cramping, spasms
- Diarrhea: urgent, frequent, even incontinence
- Constipation
- Alternating diarrhea and constipation
- Undigested food in stools
- Mucus in stools
- Gas: frequent, bothersome, noticeable

- Food avoidance
- Food craving

How to Use This Book

Trust Your Gut is organized according to the guiding CORE principles.

Part I: Center

Chapter 1 addresses the foundational concept of centering, a notion closely tied to the ancient idea that your gut is the center of your emotions, energy, and intuition. As you learn to trust your gut, you begin to realize that it is the source of healing power and can help your body/mind system to regain its balance. Proper function replaces dysfunction, and symptoms begin to go away—sometimes very quickly. You will also learn specific strategies for centering practice as well as ways to identify your own strengths and inner resources. Centering is the prerequisite for the skills that follow.

Part II: Observe

To begin the healing process, you must first become aware of yourself, your bodily sensations, and the various forces that keep you out of balance. Only then can you restore your balance. However, many gut sufferers tend to ignore their pain, mask it with medications and painkillers, or distract themselves from their physical sensations by worry and anxiety. Chapter 2, "Observe Your Gut," teaches you how to pay attention to the unpleasant sensations in your gut with greater acceptance and how to interpret them as meaningful clues about your problem. The very act of shifting the focus of your attention helps you to become more centered and calm.

Chapter 3, "Observe Your Stresses," looks at the five main sources of stress on your body/mind system: environmental, physical, emotional/spiritual, pharmaceutical, and dietary. Most gut sufferers carry a negative connotation of the word *stress* and take it to mean "your digestive problems are all in your head." Stress is actually a normal part of life, but it can directly affect your gut distress. We will guide you through a self-assessment of the five forms of stress in your life. From this will come the awareness necessary for effective and lasting change.

Chapter 4, "Observe Your Diet," presents ways to start tracking how the food you eat may be correlated to your symptoms. These results can help pinpoint foods that need to be removed from your diet.

Chapter 5, "Observe Your Sleep," examines the very close relationship between chronic gut distress and sleep problems. Abdominal pain, cramping, and diarrhea during sleep hours keep you awake, and the lack of sleep makes you more vulnerable to gut distress the next day. This chapter features a self-assessment of the most important dimensions of sleep along with specific solutions to guide you out of this vicious cycle.

Chapter 6, "Observe Your Life: Your Health and Wellness Autobiography," will teach you how to write your own medical biography. The better you know your own health backstory, the better prepared you are to write the next chapter—one featuring health. The key point of observation is that the more aware you are of your physical sensations, stresses, emotions and thoughts, and lifestyle, the more you can activate self-healing resources. Writing your medical biography is a vital tool for this healing awareness.

Part III: Restore

You need new rebalancing skills to restore harmony to the systemic dysfunctions that are causing your intestinal problems. These skills will bring lasting relief as opposed to the use of pills and pharmaceuticals that cover up the symptoms. Chapter 7, "Ecological Rebalancing for Inner Peace," and chapter 8, "Harmonizing Your External Environment," show you ways to restore your interior and exterior ecological systems. These chapters include the newest scientific findings regarding the microbiome, the trillions of microbes that coexist in our gut and elsewhere in our body. Tools to manage your interior ecology include special diets, probiotics, and supplements. Balancing your exterior ecology examines such things as the need for rest and rejuvenation as well as your relationships with work and with other people.

Chapter 9, "Neurohormonal Retraining to Rewire the Gut-Brain Connection," teaches you a powerful new technique to retrain the way your brain responds to sensations in your gut. For many gut sufferers, the very anxiety of dealing with their chronic problem conditions the brain to overreact to each little digestive gurgle and twinge. Neurohormonal Retraining short-circuits this misdirected feedback loop and helps restore the natural harmony between brain and gut. Over time, our body becomes conditioned to experiencing chronic pain, which alters and distorts the communication between the brain and the gut. You will learn how to take advantage of neuroplasticity, the amazing ability of the brain to create new nerve pathways.

Whatever the nature of your particular imbalance, certain skills are universally applicable. Chapter 10, "CORE Calming Techniques," presents additional tools for calming and balancing your body, your nervous system, and your emotions so you can heal more quickly. These skills include meditation, breathing exercises, and self-hypnosis—a scientifically proven tool for easing digestive distress. This daily regimen will help your system run more smoothly and be more responsive to the powerful techniques of Ecological Rebalancing and Neurohormonal Retraining.

Chapter 11, "Resolve Difficult Emotions and Their Physical Effects," covers the crucial but often misunderstood topic of how emotions negatively interact with gut symptoms. Stressful emotions don't cause IBS, but they certainly are a contributing factor and can worsen your symptoms. We will guide you through a self-assessment of stressful emotions and will provide specific solutions for gaining better control over them. It will also help you respond to stressful emotions as helpful signals to take better care of yourself.

Part IV: Ensure

Life never stops, and the world keeps changing. Once you have centered, observed, and restored your body/mind system, you will constantly need to adjust to unexpected obstacles and traumas in your life that may upset your balance. Chapter 12, "Maintain Your Gains and Achieve Sustainable Lifestyle Change," provides ways to handle these bumps on life's road. Sometimes people feel so good after their gut distress goes away that they become over-confident or complacent and return to their old ways of life—only to see their symptoms return with a vengeance. The elements of the CORE system need to become habits, a sustainable and natural part of your life so you can bounce back from occasional flare-ups and setbacks. Finally, we provide you with specific solutions to help you push past the normal resistance to change and awareness, so you can move ahead with your progress.

We wish you well on your journey to reclaim control over your life. We are confident that the techniques and skills you learn from this book will significantly reduce your suffering. We would also like to gently remind you that in the CORE program, your most important healing partners are your own body, mind, and spirit.

—*Greg Plotnikoff and Mark Weisberg*
Summer 2012

PART I

Center

Centering is the first step on your path to recovering a life of balance, harmony, and comfort.

Like the vertical balance of stacked stones, if we are in harmony with our environment, we are balanced. Like the horizontal balance of a teeter-totter, if we are centered, balance is nearly effortless.

Balance a pen on your finger. Notice the effort it takes to find its center and achieve balance. Notice also how effortless that balance is once you have found this center. The same is true in our life. Once we have discovered our own center, balance follows almost effortlessly.

Notice also that if the pen is not balanced on its center, it takes significant effort to make the pen appear balanced. It is possible to keep the pen horizontal, but does this feel natural, elegant, or easy?

This is where many people find themselves. Great effort goes into appearing balanced, rather than actually being balanced. This struggle takes its toll on health and well-being.

This is especially true for all people whose lives are disrupted by gastrointestinal symptoms. You, like so many others, may put great effort into appearing balanced and in control. You may use words like struggle, battle, *and* try.

Natural, elegant, *and* easy *are probably not words that come to mind. By centering, however, balance is inherent.*

Centering is the foundation and the goal for everything you will learn throughout this book. Let us begin our healing journey with centering.

1

Find Your Center

Flow with whatever may happen and let your mind be free. Stay centered by accepting whatever you are doing. This is the ultimate.

—Chuang Tzu, Chinese philosopher

Your mind and body make up an integrated system, and when it goes out of balance, you become dysfunctional. The results of this imbalance are obvious in people who perform for a living. An Olympic gymnast who is uncentered crashes on the floor. An actress forgets her lines. A juggler drops the balls. A batter can't hit the baseball. That's why such performers always prepare themselves with some sort of centering technique before the curtain rises or the first pitch is thrown. They get psyched up before the big game to keep their mind calm and focused, and their body flexible and alert. The mind and body must become one. When they stay centered, they perform perfectly—the slugger gets a hit and the gymnast gets a 10.

It's no different if you suffer from chronic intestinal distress. You have an imbalance in your body/mind system, and you can only find lasting relief by becoming centered. No one sees your problem, but you know it preoccupies you way too much of the time. Whenever you walk out the door to go to work or out on a date, you are on stage. The problem is you don't always understand how to center yourself beforehand. You lack the techniques to keep your mind sharp and your body under control. You have been given various pills and been told to relax, but it doesn't exactly work. You never stop worrying, and you never feel in charge of your life.

You need to know where you are headed—and why—before you can become centered.

Centering is the first step in our CORE program. You need to know where you are headed—and why—before you can become centered. That's why we begin the CORE program with centering.

Uncentered Kevin

It was hard for Kevin to make it through his day's work because he just couldn't concentrate. His mind was preoccupied with the dread of an imminent attack of gut pain, bloating, diarrhea, or constipation. It was hard to sit still, and if he did sit still and happened to notice a sensation in his gut, he'd get quite angry. He would urgently start strategizing how to respond to this latest attack. Should I take a laxative? Should I go for a walk? What the heck am I going to do?

He was so off center that he couldn't even focus when he was at home with his family. His anxiety and worry about his symptoms took over his entire day. All he could do was anticipate the worst.

His bloating was so bad, he bought extra pants to accommodate his expanding waistline. He had a thirty-six-inch waist, but he kept a thirty-eight- and a forty-inch pair on hand to wear on any given day, depending on his level of bloating.

Kevin increasingly saw himself as being damaged. He had no hope that his symptoms could get any better. These cumulative anxieties kept stirring up his nervous system and made the bloating and the pain even more intense. Kevin was totally off balance, but he never even thought about it that way. His physician kept giving him medications for his symptoms and his psychologist kept telling him to relax. Nothing worked, and that just made it worse.

Becoming Centered Is a Process

Telling people to relax doesn't make them relaxed, unless they already know how to relax. If an angry parent is yelling at his son's little league coach during a game and you tell him to relax, he's more likely to punch you than to mellow out. But if you ask a Buddhist monk who has practiced meditation for thirty years to relax, he could easily produce ultra-calm theta waves within a minute or two. Likewise, you can tell a professional opera singer to get centered, and she could become poised with a few deep breaths. But if you told Kevin to get centered, he'd only get more frustrated. He'd be more likely to resemble the angry parent than the Buddhist monk.

Learning how to become centered requires a change of attitude and the acquisition of new skills. It's not a mere intellectual process that only requires thinking—it's an experiential process, an activity. The Olympic gymnast may not be able to verbalize what it is to be centered, but she certainly knows how it *feels*. Being centered is a psychophysiological state—both physical and emotional. It is also embodied; you can feel it in your gut. If you keep thinking too much and a worrisome dialogue keeps replaying in your head, you're never going to finish your routine.

If you are a gut sufferer and find yourself in a hopeless dead end, the most important step on your path toward centeredness is to learn to trust your gut. This means getting a new attitude to replace the current mixture of hate and fear you have for your gut. As we mentioned earlier, ancient wisdom tells us that the gut is the seat of the emotions and the focal point of human energy. We can all learn much from this idea of the gut as a kind of second brain.

The Ancient Wisdom of the Gut as Center

Our everyday language uses phrases that depict the gut as a source of power, emotions, and intuitive intelligence. We say a person with a strong will has a lot of guts. A brave person performs gutsy actions. We praise one's intestinal fortitude. But those who show great fear and run away at the sight of danger are gutless cowards or yellow bellied. Even a slight fear such as stage fright before a performance can give you butterflies in your stomach. And when we know something through an intuitive hunch, we attribute it to our gut feelings or our gut instinct.

When you exercise or play sports, you can feel that your gut is your center of gravity. Balance is everything when you perform well. In traditional Asian medicine, the gut is the center of the body in another way: it is the source of your life energy. That center also requires a balance, because it is when our energies become imbalanced that we become ill. The gut is our battery, and we must live a lifestyle that keeps it well charged with energy. Because everything in your body/mind system depends on this energy, a lack of chi can affect your mood as well as your performance. In Japan's *kampo* tradition of medicine, the diagnosis of all illnesses begins with examining the gut.

If you've ever done yoga, Tai Chi, or any of the martial arts, you know what it is to feel that energy course through your body. It has different names—*prana* in Sanskrit, *chi* in Chinese, and *ki* in Japanese—but it all means the same thing: the vital, life-giving, and life-sustaining force necessary for health. This flow of energy from the center is the basis for success in the martial arts, Zen meditation, flower arranging, Zen archery, and every other mindful activity. Centered practitioners perform in a relaxed and effortless manner with calm and focused minds. Like the best actors and dancers, they make it look easy and natural.

Asia wasn't the only place where the gut was seen as a major center of vitality and emotion. Some translations of the Bible also depict the guts as the seat of strong emotions such as compassion, mercy, intuition, and empathy. For example, in the story about the wisdom of King Solomon, in which he proposes cutting a baby in half to solve an argument between two women who both claim a child, the Cambridge edition of the King James version says "her bowels yearned upon her son" (I Kings 3:26). In our effort to restore your faith in your gut, we are harking back to the wisdom of the ages.

CENTERED	UNCENTERED
Relaxed	Anxious
Effortless	Struggle
Focused	Scattered
Functional	Dysfunctional
Aware	Confused
In control	Out of control

The Breath Connection

Everyone knows that the gut is the center for the ingestion and digestion of the essentials for life—food and water. But the gut is also the center of our breathing apparatus. Sure, the lungs are what fill with air, but the abdominal muscles are what provide the strength of the bellows that keep us alive. If you watch a baby breathe, you will see her belly expand and deflate. That is natural deep breathing. Asian medicine acknowledged this truth by naming the energy that flows from the center after the breath. *Chi, ki,* and *prana* all literally mean "breath." Actors and singers around the world are taught to breathe from the gut. They know that you get more air that way and need to pause for a breath less often. Breathing from the chest is a human invention that takes in less air. Gut breathing is deep breathing, while chest breathing is shallow.

Breathing is one of the few bodily processes that run automatically when we are not paying attention, but yet we can take control of our breath when we want to. This is useful because our rate of breathing correlates directly to our state of mind. Deep breathing makes us calmer and more centered, but when we are uncentered, confused, and anxious, our breath rate and pulse both become more rapid. This breath connection is evident in the case of Carol.

Carol Gets Calmed

Carol was a senior executive who suffered from a long list of medical conditions—constipation, bloating, fatigue, poor concentration, and much more. She sought the advice of many doctors but to no avail. She felt hopeless, and she blamed herself for her condition. "I am a mess. My gut is a mess," she said. "After I eat, I bloat so much, I look six months pregnant. I am so sensitive to everything—if I could just get calmed!"

She was finally referred to Dr. Plotnikoff, who had Carol keep track of her diet and symptoms for two weeks. When she began to read her notes to him, she was so scattered and nonlinear that her efforts to please even sent Dr. Plotnikoff off center. He was too distracted by her frenzied effort to hear what she was trying to say.

After ten minutes, he realized Carol was so agitated that he needed to interrupt. He sensed that she needed to focus. He moved on to the physical exam and told her he wanted to check her pulse. "I took her right hand in mine and placed my left hand over her right wrist to feel her pulse. I noticed that she closed her eyes. I felt her pulse for one minute. Her hand was not cool or damp, as I had expected. Her pulse was a very reasonable seventy-four beats per minute. I switched to her left hand for another thirty seconds."

The energy in the room changed significantly with that simple act of checking her pulse for a minute and a half. They were both able to center. He asked what she was feeling, and Carol reported a sense of calmness and hope, of actually feeling better. He then led her in some breathing exercises focused on breathing into her center. She left the clinic having discovered one approach for centering and grounding herself.

The Emerging Science of the Gut: The Intestinal Brain

Western science has increasingly come to consider the gut as much more than just a digestive tract. In the last twenty years, scientists have researched the neuralhormonal complexity of the gut, and more and more are now referring to it as the second brain. The intestinal nervous system (or enteric nervous system) is composed of a cluster of more than 100 million neurons. It has receptors for more than thirty neurotransmitters—the hormones such as epinephrine, serotonin, and dopamine that allow a neuron to send a message to another neuron. In fact, more than 90 percent of the serotonin receptors and more than 50 percent of the dopamine receptors are in the gut.

Of course, the brain in our head is vastly more complex and has a thousand times more neurons than the intestinal brain. However, like the main brain, the intestinal brain receives, organizes, and transmits information. That means that both brains allow rapid and coordinated responses to changes in the environment, and both brains can regulate our internal organs.

The intestinal brain has two main connections to the main brain: a calming route along the vagus nerve and an energizing route along the spinal cord. Both connections operate automatically as part of the autonomic nervous system. When your body/mind is balanced and centered, the calming and energizing parts of your nervous system are likewise balanced. They are complementary. But when these two systems are out of balance, the result is often major intestinal problems like pain, bloating, diarrhea, or constipation.

For chronic intestinal problems in which all the life-threatening diseases and maladies have been ruled out, one major cause of dysfunction is that your two brains have somehow gotten their wires crossed. They have become conditioned—just like Pavlov's dog—to react to a threat when no threat exists. That's why it can't be fixed by a pill. The problem is not a disease but rather something closer to a computer virus. It is a system gone awry. The problem is not in your head; it's in your wiring.

When your body/mind is balanced and centered, the calming and energizing parts of your nervous system are likewise balanced.

Imagine a feedback loop that is out of control—such as a sound system in an auditorium when someone talks into the microphone and you hear a squealing feedback sound. The problem in this loop is that the microphone is oversensitive and picks up not just the normal voice but also the amplified voice over a loudspeaker. Then the microphone sends the amplified voice back through the amplifier and out the speaker again, only louder and more shrill than ever. In a fraction of a second, the shrieking sound gets so loud, it hurts your ears. The speaker has to stop because nobody can hear her anyway, and then you have to turn down the microphone or move the loudspeaker farther away to interrupt the feedback loop.

In the case of an attack of digestive distress, instead of an oversensitive microphone you have a hypersensitized amygdala—a primitive part of the main brain that decides whether a threat exists. It can take a small, harmless sensation and encode it as threatening. This sends a danger signal to the gut, which reacts by tensing up and causing distress. The intestinal brain sends these amplified distress signals back to the amygdala, which totally freaks out and sends more emergency signals back to the gut, so then the gut goes bonkers as well. The feedback loop has gone berserk and keeps accelerating, but instead of a terrible noise in an auditorium, you get awful pain and distress in your gut.

Sally Sees a Tums

Sally was a young professional who suffered from IBS and had recently gone through a painful diarrhea and constipation cycle. She was on her way to a date and stopped in a convenience store for lip gloss. While there, she saw a shelf of Tums and other digestive remedies. Almost instantly she felt a minor rumble in her abdomen. What just happened?

The main brain and the intestinal brain just had a little scene together. The main brain took in visual input of Tums, which sparked memories of recent diarrhea and constipation, and automatically assigned an emotional evaluation of threat. The oversensitized amygdala exaggerated the severity of symptoms and sent an alarm message via the spinal cord to the intestinal brain, which activated her gut. If the threat is seen as a crisis, the system releases stress hormones such as cortisol or adrenaline—which cause a series of reactions including tightening of the gut muscles, resulting in pain, bloating, cramping, and more. Sally ended up canceling her date—not because she was sick but because she saw a Tums and that set off a feedback loop gone bad.

Neurohormonal Retraining

The good news for Sally and all gut sufferers is that there is an adjustable link in this automatic chain of events. The part of the brain that decides whether a threat exists, the amygdala, is retrainable. On the negative side, the amygdala can be falsely conditioned to arouse a fear response when there is no actual danger, thus setting off a feedback loop gone awry. But on the positive side, the amygdala is the loophole in the main-brain/intestinal-brain circuitry that provides an opening to fix the erroneous programming. The process of fixing this feedback loop is called Neurohormonal Retraining, a key skill you will learn in this book.

Ecological Rebalancing

Because everything is connected, a variety of imbalances in your body/mind system can have negative effects on the function of your gut. Your connections outside your body comprise your exterior ecology—everything from your personal relationships and home life to your workplace and environmental surroundings. Your interior ecology includes the food you eat, the levels of vitamins and minerals in your system, and the health of your microbiome—the 100 trillion microbes that live inside you. These are the bacteria that help you digest food, strengthen your immune system, and keep you in a good mood. While you may not find it amusing that several pounds of microbes are dwelling in your gut—far outnumbering your human body cells—if your microbiome is imbalanced, it could be a cause of your gut distress. Throughout this book you will discover techniques for balancing your inner and outer ecological systems.

How to Start the Process of Getting Centered: A Three-Step Exercise

Now that you know the importance of centering for your gut health, it's time to begin putting it into practice. Here are a few steps you can take right now to begin the process of centering and the journey of your CORE healing:

1. Get grounded.

2. Identify your strengths.

3. Set your intention.

Get Grounded

Grounded is a common term used to describe being calm, centered, relaxed, and focused. Yet most people don't know how to deliberately achieve this experience. Here is an exercise to help you get grounded:

1. Sit in a comfortable chair.

2. Take slow, easy breaths for 30 to 60 seconds. Breathe in through your nose, into your center, and slowly exhale through your mouth.

3. Pay close attention to your senses (visual, auditory, sensory/ kinesthetic) in your body and what they take in. Spend 1 to 2 minutes on each of your senses.

For example, start with your vision. Simply sit in the chair, look around the room, noting any visual details you can see. What do you notice? You may see a plant sitting on a table. A piece of artwork on the wall. The clock. A couple of table lamps. The tiles in the ceiling. And so on. Just notice how many visual details you see, without analyzing or making any judgments.

After a minute or two, switch to auditory (sound) awareness. Do the same thing. Perhaps you'll notice the sound of the ventilation system, cars driving by, a muffled voice in the next room, and so on. Next, switch to sensory/kinesthetic awareness. Notice the sensation of the bottom of your thighs being supported by the chair. Feel the temperature of the air on your skin, the weight of your jewelry, or the tightness of your clothes. Don't judge or analyze; just feel.

That's all there is to it. What do you notice? Among other things, you'll feel calmer, your mind will be quieter, and your attention will land more in the present moment and not stuck in the past or the future. This comfortable feeling is good preparation for the next step in the centering process. Remember how good it feels to be grounded.

Identify Your Strengths

Think of a time in the past when you encountered a situation that was challenging or downright difficult, but you ended up successfully achieving your goal. Take a minute to close your eyes and breathe slowly and gently. As you calm down, it becomes easier to identify your strengths. Let yourself drift back to an earlier time in your life when you overcame a big challenge.

Remember as many details as possible about the obstacles you faced. Now, review everything you did to attain your goal. As you reflect, try to identify the skills and abilities, both mental and physical, that helped you succeed. Don't be shy or modest. Feel free to acknowledge all the positive attributes that you were able to bring to bear on the problem you faced. If it is hard for you to see your own strengths, ask your friends, family, or coworkers. They are in a position to be open and honest about your strengths.

There seems to be a direct correlation between being centered and feeling confident about your strengths. If you are ungrounded, you may be temporarily blinded to your strengths. Some people perceive themselves as so weak that the very mention of personal strength evokes grief and shame. On the contrary, the more grounded you become, the greater your personal power becomes. The fact is that we all have strengths, and when we are calm, we can call upon those strengths more easily and build upon them to achieve our goals. For example, some of Dr. Weisberg's patients become so grounded and adept at self-hypnosis that they can undergo surgery using only a very small amount of anesthesia. Their centered concentration is so focused that they can voluntarily shut off the feelings in the nerves near the site of the operation.

As you review your personal resources, have something to write with nearby and list all the strengths and skills you used.

Take that list and keep it in a safe place. Refer back to this list frequently as you participate in the CORE program, as you'll be using these same strengths and skills.

The following example illustrates how you can use this valuable exercise to identify your strengths and resources for healing.

Jim Identifies His Strengths

Jim was a carpenter in his thirties who suffered from chronic indigestion, gas, bloating, and abdominal pain. Dr. Weisberg invited him to remember a difficult situation when he ended up being successful so we could identify his strengths. He had no trouble coming up with an example.

Jim had been overweight all his life, but he just accepted it because his entire family was overweight. However, when Jim turned twenty-five, his new primary care physician, Dr. Taylor, changed his attitude. He convinced Jim to act by gently explaining the various health risks of being forty-five pounds overweight: a greater chance of cancer, heart disease, diabetes, immune system problems, and so on. Jim accepted Dr. Taylor's recommendations for dietary changes—including exercise and Weight Watchers— and kept at it until he lost thirty-five pounds. What's more, he's kept that extra weight off ever since.

Jim wrote down his recollections of this challenge and the inner and outer strengths that helped him reach his goal:

> *Once I realized I needed to lose weight, I decided to stick with this plan [decisiveness]. I stayed with it even when I didn't want to [determination, persistence]. I felt good about the helpers I had chosen to assist me: Dr Taylor, a nutritionist, personal trainer, and psychologist [faith and trust in my team]. I was able to look rationally at the problem, study the research, and understand the importance of sticking with this regimen [intelligence]. As I got used to the new diet, I'd make jokes about how*

I didn't even miss what used to be my favorite foods, like ice cream, bread, potato chips, and beer [sense of humor]. My girlfriend, Jill, who is now my wife, really encouraged me and cheered me on, as did my fellow members of Weight Watchers [relationship and social support]. I also realized that this was a process that was going to take some time and that the changes weren't going to take place overnight [patience, good perspective of time and change]. I also found new and interesting ways to make exercise more enticing, rewarding myself different ways for each week that I stuck with my workout schedule [creativity, giving himself positive reinforcement].

Afterward, Dr. Weisberg recited the list of strengths he found in Jim's story.

"I've known on some level that I've accomplished things before," Jim replied, "but it's really helpful to list my strengths so directly. I'm going to be even more able to call on great resources to reach my goals this time!"

Set Your Intention

Once you have identified your strengths and resources, it's time to set your intention for healing. Many successful performers, athletes, and businesspeople use visualization and intention setting as part of their success regimen. They know that if they can set their intention and picture it, they can achieve it!

Here's how you do it. Have some paper and pen nearby. Sit in the same comfortable chair where you just got grounded. Close your eyes and take a few slow, deep breaths. Now, ask yourself: how do I look and feel in the future when I'm healed from my gut distress?

Many people never even imagine this because they think it's impossible or they fear disappointment. But it is possible, and you can do it. Picture a time in the future when your gut feels

Picture a time in the future when your gut feels better.

better. What do you look like in this future image? What is your facial expression? How does your body feel when you finally have the relief you desire? What are your emotions? What activities do you see yourself enjoying? What do you notice about your level of energy when you feel this way? Pay very close attention to the details of these images. When you've finished visualizing, write down what you just pictured. You are starting the centering process, the first step in the CORE program for healing. You have already made this outcome more likely by having the courage to picture your successful outcome.

The Centered Kevin

After successfully emerging from the CORE program, Kevin was calmer and happier. He was less exhausted and had more energy. He was more hopeful about himself and his future. The bloating was significantly reduced, and he was able to stay with just one size of pants. He took great pleasure in the activities of daily living. He felt a sense of pride and accomplishment that he had learned how to regain control of his life.

"You know," said Kevin, "I always used to hate my own guts, but now I can find that when I have a gut sensation, I can listen to it without being afraid. Sometimes my rumbling is simply telling me that I am hungry. Sometimes cramping is a sign that I ate some food that was not good for me. If my bowel movements are a little bit different, I may have been working too many hours. Now I feel more connected and more able to enjoy little things around me. I can sit and listen to music, and I hear enjoyable things in it that I never noticed before. I'm no longer preoccupied with fighting my gut."

Back to the Center

So what do we mean when we say trusting your gut is the key to becoming centered? Our patients who have successfully gone through the CORE program have learned to respond to their intestinal sensations in such a way that they no longer experience them as threats. The sensations have become friendly sources of information, helpful messages rather than something to fear. These people have broken out of the vicious circle of pain and distress, bloating, diarrhea, and constipation. They have learned to trust their gut, and they have changed their attitude, allowing their body/mind system to become more centered. These people are now more aware and observant of many aspects of life. Because they have changed their habits, their intestinal suffering is greatly diminished and often eliminated. You can do the same.

The Five Most Important Supplements to Support Centering

Centering may be unnecessarily difficult if our nutritional status is compromised. To ensure a strong and healthy metabolic core, we recommend supplementing your diet with the following:

1. Vitamin D (also called vitamin D_3): This is a crucial hormone that turns on or off more than 2,000 key genes in our body. There are vitamin D receptors on every cell of our body, including throughout the brain and the intestines. Low levels of vitamin D in the blood stream are associated with increased risk of an incredible array of diseases as well as severity of multiple others including nonspecific musculoskeletal pain, muscular weakness, asthma, allergies and autoimmune disease. Get tested and take supplements as need. Aim for a blood level of 40–60 ng/ml.

2. A good multivitamin: For us, a good vitamin is absorbable, gentle, and made of high quality ingredients. Watch out for vitamin E as DL-alpha tocopherol. If you see this *DL* on the label, think "don't like." Choose mixed tocopherols (and tocotrienols, if possible) that include alpha, beta, delta, and gamma forms.

3. Fish oil or krill oil: These supplements are rich in omega-3 fatty acids (EPA and DHA). Low levels of omega-3 are associated with gut distress including pain and cramping. A fish or krill oil supplementation blocks excessive inflammation associated with the high levels of omega-6 fatty acids found in most modern diets. Our recommendation is 1,000 milligrams a day of EPA and DHA. Check the label and add these together to understand how many capsules a day to take. Here's a secret: refrigerate to prevent fish burps.

4. Magnesium in absorbable forms (citrate, glycinate, taurate, or malate): Take 400–800 milligrams a day to support over three hundred key chemical reactions in the body. These include reactions relevant to energy, mood, memory, sleep, and general oomph power. Magnesium oxide at a dose of 400 milligrams relieves constipation. This can be taken along with the more absorbable forms of magnesium. Caution: All magnesium forms can cause loose bowel movements. Start with a low dose, approximately 120–200 milligrams a day and slowly increase as tolerated.

5. Probiotics: These will be described in great detail later. Choose a product that provides at least 20 billion CFUs (colony-forming units) and multiple lactobacillus and bifidobacterial species. Take with cool, unchlorinated water at least thirty minutes away from warm food or drink.

PART II

Observe

Awareness, in and of itself, is transformative.

—Ken Wilber, American psychologist and philosopher

Centering prepares you to turn your attention and awareness to your body and your life. This section guides those observations so that you gain a new understanding of yourself, your gut, and your habits. By looking at yourself objectively, you can begin to see the interconnectedness of everything you do and how it affects your health. Observing your gut, your stresses, your diet, and your sleep gives you clues about the nature of the imbalances in your life that culminate as intestinal distress.

The idea that healing comes from finding a proper balance in your life is nothing new. Hippocrates, the ancient Greek known as the father of Western medicine, wrote that health depends on a balance between diet and other aspects of life, including interaction with the environment. By diet, he didn't mean a "weight-loss program." The original meaning of diet *included all aspects of life under human control—one's relationship with air and water, food and drink, motion and rest, sleep and wakefulness, and feelings and passions as well as one's elimination of bodily waste.*

Our culture seems to have lost the sense for the importance of nutrition and lifestyle. However, these are central to the CORE program, which, like ancient Greek medicine, emphasizes observation for self-knowledge. From awareness, you can turn to self-healing in Part III, "Restore" and Part IV, "Ensure."

Part II consists of these chapters:

2

Observe Your Gut

You ever notice when your stomach begins a conversation with you, you're always in a silent place like the dentist's office . . . it even starts talking in complete sentences.

—George Carlin, comedian

When your gut starts talking, you need to listen before you react. The very act of paying attention to what your gut is saying helps to break the automatic neurohormonal feedback loop of gut distress. When a person with chronic gut issues is running on automatic, each intestinal gurgle or twinge sets off alarm bells in the brain. When the main brain is in alarm mode, stress hormones are sent to the gut, prompting all the symptoms you dread: cramps, pain, bloating, and diarrhea. Now your gut is not just talking to you, it is shouting—but you still don't understand. It's like a foreigner who is trying to tell you something important, but he doesn't speak your language. He tries speaking louder as if that will help you understand. It doesn't work to ignore him; he just gets more agitated because he truly has something important to tell you, but you don't know how to listen!

As a gut sufferer, you are in this very position. Your gut is trying to speak, but you can't understand what it's trying to say; the messages are garbled. But your gut keeps trying to be heard. Each cry is louder and more urgent than the last. Finally the messages become too obnoxious and painful to bear. You become frustrated and anxious and just want to stop all the belly aching. You look for pills and other remedies to silence the beast within you. You just want it to shut up and quit bothering you.

However, you have learned by now that the belly aching never stops because your gut insists on being heard. Ignoring the pain doesn't make it go away. Instead, we want you to learn the language of the gut. Gut distress symptoms are like emails tagged "urgent" that your body sends to your conscious mind.

Learn the language of the gut.

Bill Learns to Listen

Bill was a manufacturer's rep in his twenties who suffered from abdominal cramping, bloating, pain, and constipation for at least eight years. His symptoms made his work very difficult because he was on the road several days a week traveling over a five-state area. The long hours in his car, the quick stops for fast food, and the uncomfortable motel beds all made his intestinal discomfort worse.

Like many gut sufferers, Bill had run the gamut of doctors and clinics trying to find answers and relief for his distress. After taking all the tests and seeing all the specialists, Bill was told that he had IBS. The physicians tried to help, but ultimately they encouraged him to learn to live with it.

This advice backfired. Bill's frustration only made his symptoms worse. He'd cancel sales calls and break dates with his girlfriend. He just felt too lousy. Bill was at the end of his rope. Finally he went to Dr. Weisberg, who asked him, "How often do you actually observe your gut distress?"

Bill said, "Are you kidding me? I observe this distress 24/7! It's always with me, never leaves me alone!"

"Yes," Dr. Weisberg responded, "but how much of that time do you actually just sit and notice the sensations? If you're like most of our patients, when you feel a spike of pain, bloating, or pressure,

you probably notice the feeling of it for about five seconds. This annoys you and makes you anxious, and then you probably start thinking things like, Oh no, not this again. How long is it going to last this time? What happens if the pain and bloating interfere with my sleep tonight?

Bill thought about it for a minute, smiled slightly, and said, "Now that you mention it, I think you're right! How did you know?"

"These are very natural responses. I see this all the time."

Dr. Weisberg told Bill to point to where he felt the pain and bloating. Bill pointed to an area about six inches long in the center of his lower abdomen. He rated the intensity of the discomfort at five out of ten.

"Good," *said Dr. Weisberg.* "Now focus on the sensations in that area you just showed me. Imagine you're just going to pull up a chair and sit right next to those sensations exactly the way they are, without changing them or trying to make them relax in any way."

Bill looked puzzled. "But I do want to change them—I want those bad feelings to go away as quickly as possible."

"Of course you do, but the way that you've been trying to make them go away hasn't been working very well, has it? So let's experiment on a new way of working with all of this, okay?"

Bill nodded and settled more fully into the chair. As he sat with the sensations and focused on them, he looked puzzled, almost surprised, as if he had never really paid such close attention to his sensations before. "It's interesting," *he said.* "There's actually a lot of movement going on there. The left side of it feels now like a slight throbbing sensation, and the right side feels kind of dull."

"Very good," *Dr. Weisberg replied.* "Just make room for the left side of it to throb and the right side to feel the way it does. All you have to do is keep observing it and make room for it to do whatever it's doing."

Bill watched his intestinal show unfold for six or seven minutes and then looked up with a pleasant, yet surprised, expression. "It's really something!" *he exclaimed.* "The left side throbbed more, *and then the throbbing moved over to the other side. The area of bloating seemed to get bigger, and then smaller. I was really amazed to see how much everything was moving and changing. That fact was comforting to me because it meant my gut wasn't stuck and could actually change. It seemed like the more I made room for all those different sensations to be there, the less noticeable they became."*

Plus the pain decreased. "It was at a five, but I can barely notice it now—I'd say it's a one out of ten. I didn't think it was possible."

Bill learned to observe his gut on his own, and when he returned for the second appointment ten days later, he was very pleased to report that the frequency and intensity of the pain in his gut was noticeably reduced. His third appointment became his last, because the symptoms hardly bothered him at all anymore.

Take Action

Not all patients find relief from their symptoms as quickly as Bill did. But practically all of our patients do experience relief and an improved sense of control over their symptoms by learning to notice and observe the sensations in this passive way. Ironically, they're learning to control their symptoms more by not trying to control their symptoms at all.

As a little experiment, just take a minute or two right now to focus on a sensation somewhere in your stomach or abdomen. Just imagine that you're going to pull a chair up to it and simply observe it. Don't try to change it, fight it, or make it relax. Just let that sensation in your abdomen be there exactly the way it is. If it stays the same, that's fine. If it moves, changes location or intensity, that's fine, too. Just be curious to notice what happens without you trying to change it in any way. Afterward, jot down on a piece of paper what you observed.

As you practice this skill and gain experience with it, your brain and nervous system will learn that these sensations are not threats. You will begin to feel the sensations diminish in intensity.

What's more, you'll find that the anxiety, dread, and muscle tension associated with those sensations will also diminish. You will feel better and calmer, and you can then begin the process of actually *listening* to what the sensations are trying to convey.

You may be put off by this whole idea. You might be thinking, *Why on earth would I want to make room for and accept sensations that I hate so much? I want to fight that sensation with everything I've got!* This is a common reaction to something that seems counterintuitive. Skepticism is normal. As one of our patients said at this stage, "No offense, Doc, but are you crazy? This pain in my gut has taken over my life in every way. The last thing I want to do is pay any more attention to it. I'd rather stick my finger in an electric socket."

We are biologically predisposed to be alarmed when we feel something painful or unusual in our body, especially when it's new. This is meant to protect us from harm. If you accidentally put your hand over a flame, that natural alarm activates a signal in your brain to pull your hand away as quickly as possible. You don't just watch it. If you suddenly notice pain or pressure in your chest with pain radiating down your left arm, you should not simply make room for that sensation. Instead, start calling 911! Similarly, if you experience *new* abdominal symptoms of pain, pressure, bloating, and diarrhea *and have not seen a physician,* you should definitely call now for an appointment.

However, if your symptoms have become chronic—and you have already been medically examined and tested—then the alarm function is not needed. It's a false alarm. The mix-up is that your more primitive, reflex-level limbic brain never got the memo. Unfortunately, most people still react with alarm, anxiety, or panic when they feel that all too familiar discomfort in their gut. Calmly observing those sensations helps to short-circuit the primitive brain.

At first, it may feel awkward or scary to observe your gut symptoms with a sense of passive curiosity. But we want to assure you that it is perfectly safe. We will be with you through this process, step-by-step, to help you master these new empowering skills. Over time, it will help you feel relief and gain more control over your symptoms.

Why Fighting Symptoms Doesn't Work

When you go on the offensive and try to fight your symptoms, it only makes them stronger. Try this: whatever you do, do not think of an elephant. What happens? The harder you try not to think of it, the elephant's image becomes even more prominent.

That's exactly what happens when you try the same game with a more emotionally charged topic: whatever you do, do not think of your gut distress. Instead of not thinking about your gut, your reactions become all the more amplified and disturbing. The harder you try to not think about the symptom, the more powerful the image. And the more powerful the image, the more likely you are to assign emotional value to it. Your body then responds automatically, unleashing multiple reactions, including the release of stress hormones and other factors that perpetuate or worsen gastrointestinal distress.

Say you get a burning pain in your upper abdomen every time you eat pizza. It's nothing new—a chronic problem. You've seen your doctor to rule out physical pathology. As you feel that all too frequent sensation, your body sends out a cascade of emotional and physiological processes in response.

First, your sympathetic nervous system shouts out for the release of stress hormones such as adrenaline (also called epinephrine) and cortisol as your primitive limbic brain mistakenly encodes the pain as a threat. Because emotions often express themselves in physical ways, when you then try to *not* feel this unwanted burning pain, you instinctively tighten the muscles and other tissue in the area where it hurts—and other areas of your body as well, perhaps far removed from the source of the pain. All of this makes the pain worse, sometimes even spreading to other parts of the body.

At the same time, conditioned learning occurs in your limbic brain; the next time you eat pizza and get that familiar burning pain, you're just a little more likely to automatically react with anxiety, anger, frustration, or despair. It's a difficult cycle, one that increases your suffering.

Avoidance Increases Anxiety

In your body's understanding, the twinge of pain in your gut has taken on a whole new meaning. Instead of merely being useful information, this sensation has become mislabeled as a serious danger to be avoided. Yet this false threat has the power to take away simple everyday pleasures like pizza parties.

Susan was learning how to go spelunking in a practice cave with narrow, hard-to-navigate crawl spaces. At one point, she got stuck in a particularly narrow space and panicked. "Oh my God, I'm stuck," she shrieked. "No, you're not," called back her instructor. "Just relax, Susan. You are *never* stuck. If you tense up, your body will naturally get bigger and you'll be even more stuck. If you relax, your muscles will go loose and your body will naturally find a way out." Tensing and resisting actually make distress worse—in spelunking and in your gut! Likewise, learning to relax your gut helps to activate your body's natural healing powers.

Tensing and resisting actually make distress worse.

Learning to Listen

Let's do a simple exercise to prepare you for listening to your gut. At this stage, don't worry about interpreting symptoms. In this exercise, you are going to observe your gut sensation and not judge or fight it.

The next time you notice a gastric symptom (gas, bloating, constipation, gurgling, churning, and so on), take one slow breath every three seconds. Allow your breathing to become easy, relaxed, and calm. Then mentally scan your gut.

Ask yourself:

- When I get this sensation, what is the thought I become aware of?
- What is the emotion I begin to feel?
- What is the sense I have of myself at this moment?

As you learn how to notice and respond to your gut's messages, you will become less alarmed and upset by your gut's sensations. Instead, you will begin to befriend them. For this self-assessment exercise as well as for activities you'll work on throughout *Trust Your Gut*, you will need a diary dedicated to tracking your work. The CORE program will help you improve your self-awareness and self-observation—on mental, emotional, and physical levels. In other words, CORE will help you change your relationship with your body and how you react to its signals.

You Can't Change It If You Don't See or Feel It

It's human to avoid anything painful or upsetting. No wonder Prilosec sells so well—it's much easier to shut down gut communication altogether than to listen to it. Millions try to avoid their gut sensations; many others just feel a sense of defeat and hopelessness. Feeling anxiety, frustration, or worry in reaction to a gut symptom need not be judged. If you attach an emotional value to this sensation, don't fight it; just notice it.

These sensations are important forms of communication, and we need to listen. Nobody appreciates it when they don't feel heard or understood. After all, many clinical studies have shown that the act of being listened to is one of the most important aspects of the healing encounter between doctor and patient. So it stands to reason that it's also important that you learn to listen to *yourself*—to the signals from your own gut. That's not so easy when you've ignored, suppressed, and rejected messages from your gut for a long time.

A Breakdown in (Gut) Communication

The gut's ability to communicate to us has been understood as normal for thousands of years all over the world. References to the importance of the feelings of the gut can be found in the Bible as well as in traditional Japanese medicine. But when did gut communication break down in the Western tradition and become seen as symptoms—something abnormal? When did we start to shut down or dismiss these crucial internal lines of communication?

To answer, we need to go all the way back to ancient Greece during the days of Plato and Aristotle. These early philosophers were part of a movement to use reason to rule our life instead of being guided by our emotions or superstitions. Plato placed such a high value on reason and mathematics that he considered them to be the highest forms of knowledge, closer to the mind of God and the higher unchanging realm of pure ideas. Emotions yielded insignificant information about the ever-changing world of everyday existence. Aristotle, one of the founders of Western science, agreed on the value of reason over emotion. Even lower in esteem than emotions was the body, which was considered a crude vessel that imprisoned the soul.

By the 1600s, many influential philosophers, most famous among them René Descartes, argued that the mind and the body were separate substances, dual realities. The mind was eternal and invisible—known only by our own consciousness. Our body was considered to be merely a complex machine. And so began medicine's distrust of the body's knowledge and the body's wisdom.

These philosophers did not have the advantage of our 21st-century technology like functional MRI scanners that allow us to actually look into the brain and literally see the chemical reactions taking place. Scientists can now observe which genes get turned on and turned off by emotional response. Human beings can now measure scientifically the interactive effects of mind and body upon each other. However, hundreds of years ago, the mind was viewed as nonmaterial, considered only within the domain of religion. Somehow the mind was less real because you can't see it or put it in a wheelbarrow. Not so with the body. This was viewed as tangible, material, real—within the domain of science. The mind was split off from the body. Though this happened hundreds of years ago, these ideas shape how we view health, illness, and our relationship with our body today.

The Industrial Revolution disconnected us even further from our natural ability to communicate with our body. New inventions allowed us to predict, measure, and control aspects of our life that were previously impossible. Manufactured chemicals were produced on a mass scale for many uses. Germ theory and vaccinations helped contain life-threatening diseases like smallpox. Scientists looked at the body as if it were only a machine, an orientation that led to many scientific advances, such as organ transplants.

This great medical progress was a double-edged sword. Although we learned that it was possible to treat and eliminate many diseases, we increasingly came to believe that painful or unusual bodily sensations should be eradicated. Accepting and listening to sensations seemed pointless. Likewise, we rejected any connection between mind and body. How could a nonmaterial thing like the mind affect a material thing like the body? Such a crazy idea violates the laws of physics. After all, no matter how hard we think, pray, or believe, we cannot get a frozen car to start on a winter morning. Mind has no effect on machines.

Ignoring Feelings

Machines have no feelings, but people do. We can be happy, sad, frustrated, joyful, and more. We can experience emotional pain and emotional pleasure. Because we are geared toward pleasure, we developed psychological defenses against emotional pain. This is a normal way to cope with the difficulties of daily life. As children, we learn these natural ways of forgetting, avoiding, denying, or minimizing painful or difficult feelings. Nevertheless, we actually do experience our emotions in our body: tightening in the forehead, tensing the jaw, heaviness in the chest, rumbling in the stomach. So when we avoid feeling something emotionally uncomfortable, we also have to fight the sensations in our body—wherever we feel emotions. That's why the word *feeling* refers to emotions as well as to physical sensations. An ignored emotion that manifests as physical discomfort is like a secret kept from the mind but not from the body.

As it turns out, ignoring our feelings is not easy and definitely not helpful!

For hundreds of years now, Western science has separated mental activity—thoughts, emotions, and reasoning—from the physical body. Over the centuries, we as individuals have learned to downplay feelings and emotions as we have played up reason. The same is true for the culture of medicine, which focuses on controlling symptoms. Emotional responses were dismissed as too subjective—or at least irrelevant. If we can't find it in your body, then it must be in your head, they thought. If it's in your head, then it becomes an issue of irrational emotions and feelings—something to be treated by a psychologist.

Quiet Cindy

Cindy lived with her boyfriend, Jim. She felt very strongly about him. She liked the closeness they shared, but she was also afraid of conflict, so she never mentioned when she was angry with him. Then Jim got into a habit of making plans to go out with other friends in the evening without letting Cindy know in advance. This bothered her, but she didn't say anything to him. In recent weeks, though, she'd been experiencing a gnawing pain in her lower abdomen every time Jim made plans without telling her.

This could be considered a secret kept from the mind but not from the body. Ultimately, through her participation in the CORE program, Cindy learned that the more she expressed her anger and other uncomfortable emotions as they emerged, the gnawing pains in her abdomen diminished. Clearly, her gut was encouraging Cindy to speak up and be less quiet.

Somatic Wisdom

We see far too many people in our clinic who feel reduced to either their body or their mind. A 360-degree view of the person as a person, not as just a mind or as just a body, has been missing from modern medicine. The CORE program for gut recovery goes much further. The CORE program will allow you to discover your inner wisdom of how emotions and physiology interact. This is called somatic wisdom—the wisdom of the body (*soma* is Greek for "body").

Nancy's Unfortunate Appetizer

Nancy was sensitive to many foods because they caused gut distress. She tried to stay vigilant about only eating foods on her safe list. One night at a party, she ate an appetizer that looked appealing to her. Five minutes later, she started to feel queasy and mildly bloated in her mid-abdomen. Being so concerned about her diet, she felt a surge of adrenaline and a feeling of anxiety.

She thought, Oh no, what if I ate something my gut can't tolerate? What happens if I have sudden diarrhea at this party and I get embarrassed? *As she worried, she felt her lower abdomen start to cramp and the queasiness increase. As her anxiety intensified and her heart rate jumped, she urgently scanned the apartment to find the bathroom. Her emotions and her physiology were intertwined. Nancy's body and mind joined in an intimate yet painful dance.*

Although our culture tends to separate the mental and emotional from the physical, our brain does not. Scientists define pain as both a sensory and emotional experience. Why? Pain signals enter our brain, where they channel through a sensory input center (the thalamus) and then are immediately evaluated by another brain structure (the amygdala) for emotional value and threat potential. Depending on this instantaneous evaluation, big physical changes can affect nerves, muscles, connective tissue, and even our immune system. Lo and behold, aching, bloating, or spasms arise in the gut.

Gladys Goes Through the Mill

Over the past twelve years, Gladys had seen more than eight doctors for her bloating, constipation, cramping, and indigestion. She'd had several complete physical examinations, numerous barium x-rays, countless blood tests, and three colonoscopies—all in the attempt to identify the problem. Doctors told her there was "nothing abnormal" on any of these tests. Identifying her problems as IBS, they tried to offer medications but ultimately encouraged her to learn to live with it. One doctor said, "There's nothing more I can do for you—maybe you would feel better if you learned to relax more."

Have you had a similar experience? Have you been told, it's just stress or it's all in your head? This last comment—it's all in your head—is another way of saying that your problems are psychological.

Gladys' next visited a psychiatrist, who evaluated her for psychological problems, which were now believed to be the source of her physical symptoms. Her physical distress was now pegged as psychosomatic. She was treated for depression and anxiety, and then sent on. Although this is sometimes effective, for most people, including Gladys, it doesn't help at all.

Sound familiar?

Gladys continued suffering the same symptoms, except now she felt a sense of hopelessness, too, because the experts she turned to couldn't help her. The result? Gladys, like so many others—perhaps like you—had fallen through the cracks of our health system. Why? Because her symptoms could not be understood as strictly due to organic physical pathology on one hand or as caused by psychological disorders on the other.

So when you experience a sensation in your stomach or abdomen, remember that it may be neither physical nor psychological but actually a combination of both! It's part of your body/mind system. Your gut sensations provide you with valuable information, but you have to learn how to listen and pay attention to them. This is at the heart of the CORE program. This valuable information is somatic wisdom, and it is available as an important healing resource.

The Media Is the Message

You probably have heard that old joke about a man going to a doctor and saying, "Doctor, it hurts when I move my arm this way." The doctor says, "Okay, don't move your arm that way." Pharmaceutical advertising turns this joke on its head, transmitting to us the exact opposite message. The message no longer says, "Don't do that." Instead it says, "Okay, you can still do that if you just take a pill first." Medications are helpful for many people, but sometimes they can also serve to mask important signals from the body.

The mass media is the way we receive commercial messages in the modern world. Whether transmitted by television, radio, print, or the Internet, messages about the gut expressed through advertising deeply affect the way people think and feel about the intimate processes of their own body. American media, in particular, is saturated with pharmaceutical ads; they're everywhere. And why? The pharmaceutical industry's very existence relies upon patients asking their doctors for "that medication I saw on TV."

What do you see when you watch a commercial for an over-the-counter or prescription gut medication? You see people with gut distress standing on the sidelines of life, not joining the party because they can't eat pizza. You hear them saying no to dinner out because they are terrified they might get a bout of diarrhea.

The hard sell? Medications put the fairness back in life. They are the great equalizer. It's unfair that your gut says, "Don't eat this food that everyone else is enjoying" or "I have to go to the bathroom even though the timing couldn't be worse." Sure, the medications let you do what everyone else is doing, but sometimes they prevent you from listening to the very important messages your gut is trying to send. Why listen to a possibly inconvenient or embarrassing message when you can take a pill and make it go away?

Medications sometimes prevent you from listening to the very important messages your gut is trying to send.

Pharmaceutical ads also reinforce the longstanding and *incorrect* belief that symptoms and sensations of the body are simply annoyances to be eradicated. They devalue the notion that the stomach irritation that flares up when you eat pizza might be useful *information about what your body needs for healing.* For example, heartburn, indigestion, and irritation are desperate signals to protect you by saying, "Pay attention to me! This food doesn't agree with you, and it's going to hurt you! Don't put us through this suffering!"

Pharmaceutical companies take a great deal of time and expense to redefine what is normal to a sufferer of gut distress. We believe your body's sensations are important messengers trying to help you heal. We believe your body's own integrity is the new normal. Drug manufacturers treat these sensations as obstacles to having a good time.

So let's assume that the guy goes ahead, takes an acid-blocking medicine, and then eats pizza until he's totally stuffed. What happens then? Probably nothing for a few minutes. But within an hour or two, he likely will feel indigestion because his body doesn't tolerate the wheat, dairy, or tomato sauce in the pizza. This disrupts the ecology of his gut. He will probably react with either constipation or diarrhea.

And this is what happens after just one episode. If he continues to use acid-blocking medicine to tolerate eating pizza despite his body's warning signs, other problems follow. For example, the acid-blocking medicines change digestion over time, because we need a certain amount of stomach acid in order to properly absorb important nutrients from our food. The result: our pizza lover ends up with an increasing cycle of pain, bloating, bowel irregularity, and incomplete digestion of his meal.

Instead of ignoring your gut, start ignoring commercials. Ads by the large pharmaceutical companies say, "Turn off your gut's messages; they are unimportant at best and harmful at worst. You can live like everyone else if you only stop listening. This pill will make your gut shut up. It will put the belly aching to rest."

The problem is that even if you ignore your gut's communication, your very wise body—your somatic wisdom—will find ways to get the message across. To find relief from the endless cycle of gut distress, it is critical to *open*—not shut off—the communication between you and your gut.

Take Action

Try this mini-exercise the next time you see a television commercial for a gut medication.

1. Watch it carefully.

2. Write down its message about your gut symptoms.

3. What is the commercial telling you about your life?

4. Does it stir up any emotions in you?

Are You Listening?

The first step in healing your gut distress and the depression and debilitation that go along with it is to rethink symptoms as messages. There is a physiological basis to understanding symptoms and sensations as messages, facilitated in part by something called the gut-brain axis. The gut-brain axis refers to the ways that the brain and central nervous system interact with the entire digestive system. It also has to do with the ways those thoughts, behaviors, and moods affect your gut—and vice versa. You'll learn more about that in chapter 9.

What does your gut's language sound like? What does it feel like? What are its cadences, rhythms, intensity, and volume?

Awareness requires slowing down, seeking silence, and then listening—really listening. You may find that your mind is restless, that you are too busy for even fifteen seconds of slow breathing each day. Perhaps silence and awareness are a bit too uncomfortable; keeping your brain too busy and distracted is how you've learned to cope—to be completely oblivious to any inner message. This is something you will overcome.

The CORE program helps you reinterpret your gut's language through the powerful practice of Neurohormonal Retraining. This process will teach you to respond differently to your sensations, and you'll begin to develop a different relationship to your body.

You will learn to sit with the here-and-now sensory experience of a symptom, without trying to fight it, change it, or make it relax. You will learn to experience the sensation exactly as it is, even as it goes through various changes in intensity, location, and duration. This may sound deceptively simple, but it is actually quite a challenge because you have been programmed to instinctively fight or avoid uncomfortable sensations. But you can change your experience of it! Focusing your attention on these sensations without fighting literally retrains the parts of your brain that learned long ago to treat the sensations as threats, setting off a cascade of stress hormones, leading to even more distress.

You will learn to sit with the here-and-now sensory experience of a symptom, without trying to fight it, change it, or make it relax.

The good news is that learning how to pay a different kind of attention to troubling abdominal sensations can lead to effective, long lasting healing as it retrains your brain-gut axis.

3

Observe Your Stresses

Reality is the leading cause of stress among those in touch with it.

—Lily Tomlin, comedian

Stress. Everyone has it, everyone jokes about it, and everyone complains about it. But no one wants to be vulnerable to it, much less overwhelmed by it.

In real life, this means that no one wants to be told that their (physical) symptoms are likely worsened by their (emotional) stress. People may too easily misinterpret such a message as, "They think I'm crazy!" or "They think it's all in my head!" Especially for people with gut distress, like Marnie, stress itself is a very stressful subject.

Marnie Denounces Stress

Marnie was a thirty-year-old buyer at an upscale department store, and she suffered from chronic constipation, throbbing pain in her entire lower abdomen, gas, and indigestion. When she came to our clinic, we gave her an initial assessment for IBS problems. When we asked her what made her symptoms worse, she replied, "I'm not really sure. I guess maybe when I eat rich food or don't drink enough water."

After asking the standard questions about her medications, other illnesses, food allergies, and possible gluten intolerance, we asked whether stress made her gut symptoms worse.

Suddenly she stiffened up and looked defensive. "These are real physical symptoms!" she exclaimed. "This question makes me so angry. I've had this pain and constipation and bloating for years now. Why is no one taking me seriously? It's not all in my head."

From a scientific point of view, stressors are *any* force that disrupts our physical and/or emotional harmony. This emphasis on *any* force is important. Stress is much, much more than an emotional issue. When people mention stress, or when we talk of stress in the clinical setting, we ask people to consider the five forms of stress:

1. Environmental (home, work, school, community)

2. Physical (illness, surgery, pregnancy, insomnia, arduous work)

3. Emotional/Spiritual (grief, loss, despair, post-traumatic stress disorder)

4. Pharmaceutical (side effects, altered absorption, nutrient depletion)

5. Dietary (allergies and adverse food reactivity, oxidative stress, nutritional insufficiencies, caloric excesses)

We highlight the five forms to emphasize that stress is much more than an adverse emotional experience. Stress is a fact of life, a normal part of experience. We can experience many stressors and not realize it. For example, honking horns, barking dogs, and ringing phones are all environmental stressors. It is easy to overlook stress in any form—physical, pharmaceutical, emotional/spiritual, and dietary—but that stress still has the power to disrupt your emotional and physiological harmony. Your job is to recognize and observe these in your own life.

Scientists now know that stress interferes with our most basic physiologic functions all the way down to the micro level of our DNA. Any of the five forms of stress can converge on the gut and disrupt its function in various ways, including:

Stress is a fact of life, a normal part of experience. Your job is to recognize and observe it.

- Diarrhea or constipation from speeding up or slowing down the gut

- Marked discomfort from increased pain sensitivity

- Food allergies, adverse food reactivity, or even colitis from reduced gut wall integrity (leaky gut)

- Inflammation from disrupted intestinal ecological balance (dysbiosis)

- Infection from stress's capacity to activate opportunistic bacteria waiting for a chance to pounce.

Even outside of the gut, stress can result in a whole range of physical symptoms such as headaches, fatigue, insomnia, brain fog, anxiety, and depression.

Because stress is such a powerful influence on our health and well-being, we do ourselves a favor by acknowledging its presence and identifying its multiple forms. This chapter will guide you in the identification and observation of the various stressors in your life. The goal is awareness, the crucial foundation for any positive and constructive improvement. With awareness of the five forms of stress in your life, you will be able to progress to the next steps of the CORE program.

Dr. Plotnikoff's Personal Anecdote

Let me share with you an embarrassing but profoundly important life lesson. See if you can identify the possible environmental, physical, and emotional/spiritual stressors present in this story.

Picture this: It's a busy morning in my inner-city primary care clinic. Patients keep getting added my already full schedule. This is a major time challenge: I have to catch a two p.m. flight, and before that I have to drop off a packet of materials at the university a few miles away. And, oh great, it's snowing. Yet again, no time for lunch today! I feel my body tense and my heart pound. I am stressed.

I somehow finish my appointments and zoom out of the clinic, rush to the university, and speed to the airport. I have my ticket and my carry-on bag, and the flight is scheduled to leave in just over an hour. I arrive and find out the parking lot is full. I am stuck in a long line of cars with suddenly grumpy drivers. I cannot move forward, cannot go in reverse, and can't move sideways. This is completely unexpected. I am really feeling time pressure and blood pressure. Now my body shifts into high-gear stress.

Fear takes hold. What if I don't make this flight?

Finally my turn comes at the toll booth. "Yes, long term is full," the parking lot attendant says. "But you can park in the short term lot for $24 a day."

"Yikes!" I exclaim, feeling it's now my turn to yell at the poor guy. "For five days? Oh no . . . "

Then he adds, "Or you can go to the econolot."

So I speed off to the econolot only to drive up and down, row after row looking for an empty spot. After multiple cruises back and forth, I finally find one. I grab my bag and run to the shuttle bus stop. And wait. And wait. And wait . . .

By this time, I am hopping up and down to shake off nervous energy. It's cold, and I really need to pee. After an incredibly long fifteen minutes, the bus finally comes. I jump on and pray that I will make my flight.

Thankfully, this was in the 1990s and many years before airport screening. I am sure that if I were in the same frantic, stressed state today, TSA would pull me aside as a high-risk suspect.

Once the shuttle bus arrives at the main terminal, I kick into Olympic athlete mode and sprint off the bus, up the stairs, and

through the minimal security check. Then I look up at the tele-
vision screen. "Gate 97 . . . oh no." Another surge of stress hor-
mones turbocharges in me, and I run at high speed through the
terminal. Somewhere around gate 45, I feel chest pain, pressure,
tightness, and discomfort. Not what anyone ever wants to feel! I
slow down a bit but push on and arrive at gate 97 just as the plane
door is closing.

"Wait!"

I charge onto the plane red-faced, sweating, wheezing, breath-
ing hard, and coughing. Looking back, I'm sure that everyone
looked at me and wondered, Does he have TB?

I find my seat near the back and immediately begin a slow,
breath-based meditation. As my body calms, the most amazing
insight comes to me. I could have dropped dead, *in the middle*
of an airport, at age thirty-five, while rushing to get to—imagine
this—a stress reduction *workshop! Hmm . . . what do you think I*
might have learned?

Yes, I could be the best meditator in the world, but that would
do nothing to prevent *such a situation. In this case, meditation*
calmed me down enough to do something even more important:
observe the multiple factors that conspired to put me in the
pressure cooker in the first place. Environmental, physical, and
emotional/spiritual stressors combined to create my painfully
embarrassing but priceless growth opportunity. Only from such
awareness can any progress come.

How Does the Stress Response Affect Your Brain and Body When You Have Gut Distress?

Stress—feeling mentally or emotionally tense, troubled, angry, or overwhelmed—can set off all sorts of gut reactions from but-terflies in the stomach to gurgling to colon spasms or urgent bowel movements. How does this happen? Scientists point to the gut-brain axis, the series of nerves that connect your stomach and intestines to your brain. Just like your heart and lungs, your intestines are partly controlled by the autonomic nervous system, the network of nerves that automatically does the body's work without any thinking required. These are among the nerves that respond to any of the five forms of stress.

Stresses of all kinds have real physical impact on the gut in three ways. The first is through what scientists call the HPA axis, which is short for the hypothalamic-pituitary-adrenal axis, the highway between two key brain structures (the hypothalamus and the pituitary glands) and the gland that sits on top of the kidneys (the adrenal gland). The key players here are the hormones known as CRF, ACTH, and cortisol, which we describe in the story below.

Janet's Cookie Causes Cortisol

Janet worked in an insurance company in the city. On the way home one day, she stopped at a gardening store to pick up some supplies so she could work in her yard that evening. They were serving cookies and cold drinks to shoppers, so Janet grabbed a cookie. Within a minute, she felt a slight gurgling sensation in her lower right gut, nothing significant for the average person. However, she had become hypersensitized to those feelings.

Janet was an IBS sufferer whose cramping, bloating, and constipation caused her to go up to four days without a bowel movement. So when she noticed minor fluctuations, she worried about having more symptoms. This kind of worry registers on the cerebral cortex, which sends electrical impulses down to the hypothalamus in the primitive limbic brain. The hypothalamus notes this threat message and releases CRF (corticotropin-releasing factor family of hormones), which finds its way to receptors on the pituitary gland, the master gland of the hormonal system. Informed of a threat, the pituitary gland sends out another hormone, ACTH (adrenocorticotropic hormone), to warn the adrenal gland. That triggers the adrenal glands to ring the alarm by sending out cortisol, a stress hormone.

The CRF and the cortisol in Janet's system started interfering with her digestive tract, and what started as a normal sensation morphed into heightened gut distress. She began to feel bloated, and the anxiety over the physical sensations caused by the rapid delivery of hormones in her system started the alarm/stress cycle all over again.

CRF by itself has multiple potent effects on the gut including inflammation, increased gut permeability (leaky gut), increased perception of gut pain, and altered gut movement. In addition to this, when cortisol is secreted over a sustained period, it can also interfere with digestion, increase inflammation, reduce immune responses, and lead to an unpleasant experience of agitation, anxiety, and physical discomfort. Cortisol has also been linked to an increase in belly fat—as if the body's response to a threat is to protect itself against starvation by holding on to a reserve store of calories—yet another bad side effect of the brain/gut distress cycle.

The second way that the five forms of stress physically impact the gut is via different nerves from the brain and spinal cord that activate a different portion of the adrenal gland. Stimulation by this route produces adrenaline, the potent rapid response hormone that powers people through the toughest of situations. Tales of ninety-seven-pound people lifting two-thousand-pound cars to save a person are stories of adrenaline in action. This is a great hormone to have when needed for fight-or-flight situations. Adrenaline clearly increases gut nerve excitability and decreases pain thresholds. The challenge is to ensure that it is not released unless really needed. Neurohormonal Retraining is one of the important methods we will teach you to reduce false alarms in this fight-or-flight system.

The third way the five forms of stress physically impact our gut is their effect on our gut ecology, specifically the 10 to 100 trillion bacteria that inhabit our gut and work with us and for us. Any form of stress can disrupt a normally harmonious ecology. Stresses such as infection, antibiotics, adverse food reactivity as well as prolonged elevations of CRF, cortisol, and adrenaline can throw the gut ecology off balance. This results in all forms of gut distress—increased pain sensitivity, bloating, cramping, diarrhea, constipation, and so on. Actions you can take to restore and ensure a harmonious relationship with your bacteria are part of what we call Ecological Rebalancing.

One main point of *Trust Your Gut* is that your digestive system is very sensitive to the hormones produced in response to any of the five forms of stress. Healing from gut distress requires subduing these stressed reactions, and the way to begin is by centering and observing—the foundation of the CORE program. These skills prepare you for the third component, restoring balance.

Here's how Janet would react after learning these skills.

Janet Cancels Cortisol

After a few bites of her cookie, Janet noticed a slight gurgling in her gut. The sensation brought to mind her recent bout with cramping and discomfort. That sent signals to her cerebral cortex, and she began to feel dizzy and off center. But Janet had learned CORE quieting techniques, so she could stop and consciously explain to herself that this feeling was natural. Then she mindfully remembered to breathe deeply, relaxing her large skeletal muscle system. Her primitive limbic brain, which is always ready to gas up her system with cortisol, instead received new information. Her deep, relaxed breathing was taken as an all-clear signal, and the false alarm was switched off. The stress hormones were held back. The gut received the all clear and settled into its happy task of digesting a cookie, undisturbed by CRF, cortisol, adrenaline, or further alarming messages.

Guide to Observing
the Five Dimensions of Stress

Stress is both a cause of trouble and the result of trouble. Stress is certainly a problem in which biological, physical, psychological, social, and spiritual factors interact. No matter the form of stress, the body reacts the same way: releasing CRF, cortisol, and adrenaline.

As the first step in empowering you to change your bodily reactions to stress, we ask you to do a self-assessment, to observe and notice what kinds of stressors exist for you in each of the five domains. The goal is awareness. As we remind you throughout the book, you can't change it if you don't see it or feel it or know about it.

In the airport story, the key to empowerment and self-healing is self-awareness. Your insights from centering and observing are the foundation for restoring and ensuring your gut's harmony.

Some stressors in life are obvious. Time pressure is well understood. And yes, if you've been through a divorce, lost your job, or just returned from serving in Afghanistan, you don't need a detective to identify your stressors. The purpose of centering and then observing the five forms of stress in your life is to increase your awareness of factors that may be quite powerful but easily overlooked. Our brain is wired so that we pay attention to new things and tune out things that are around us all the time. Plus, much of our life is filled with distractions—especially things like TV, movies, and the Internet—that can blind us to everyday sources of stress. For this exercise, we want you to take a new look at your life with a fresh set of eyes. For this reason, the first step is about centering and finding a place of deep calm within yourself. From this, you can then observe your life with more clarity.

When you're centered, tune in to your inner stress-o-meter and listen to what sets it off. With a relaxed mind-set, you can feel where the tension is

Take a new look at your life with a fresh set of eyes.

coming from. The five forms of stress clearly overlap each other, and one insight can be found in multiple categories. There are no right or wrong answers here. The five categories are merely prompters to encourage a complete review of your life.

As you examine each form of stress, also consider what resources or anti-stress elements are present in your life. If you would like, keep a running list for each of the forms of stress: stressors on the left, anti-stressors on the right. The point is to not only achieve clarity about your stressors but to also understand your easily overlooked resources and strengths.

Environmental Stress

Where and with whom you spend your time are crucial and easily identifiable elements of environmental stress. Think about what is going on at home, work, or school and in your community. Consider the physical aspects of these environments: sights, sounds, and smells, for example. Also look into the social aspects of these environments, especially the most important relationships in your life: family, friends, coworkers, and neighbors.

For most people, the biggest source of stress is the workplace. In 2009, 69 percent of employees reported that work is a primary source of stress, and 51 percent reported that they were less productive because of it. And, of course, family dynamics are also major stressors, especially with numerous opportunities for miscommunication, resentment, and other forms of tension between members. Are there opportunities for improved communication at work and at home?

Many people neglect engaging their senses in dance, music, poetry, or other forms of art. Are you at risk for not expressing your soul through dancing or singing or writing?

Also easily overlooked is the risk for nature-deficit disorder. How much sunlight is in your life? How far do you have to walk, bike, or drive to find some green space? Is the winter too long and too dark where you live? (The importance of light and day/night rhythms is discussed in more detail in chapter 5.)

As you review the possible stressors in your environment—the places and situations that make you feel uncomfortable—also keep an eye out for the aspects of your home or workplace that make you feel good. Make note of these and increase them in your life, if you can.

Physical Stress

Physical stresses such as acute or chronic illness, surgery, pregnancy, or arduous work are easily understood. Additional stressors such as dehydration, obesity, or insomnia are also easily recognized. Not so frequently seen are the physical stressors that come from insufficient movement or the accumulation of muscle tension. Compared to the lives of our grandparents, we are much more likely to sit during most of the day.

One way to observe your sources of physical stress is to keep track of what you do with your body all day long. How much time do you spend each day in front of a screen relative to other activities in your life? How much time do you spend sitting compared to standing or walking?

There are other clues to physical stressors, too. Do you experience neck, shoulder, and back pain on a regular basis? Perhaps the poor ergonomic position of your desk and chair is aggravating these aches and pains.

Aptly enough, the original meaning of the word *exercise* is to "free from restraint." Early farmers first used the word to describe the action of letting the cows out of the barn to run free in the pasture. Likewise, a wide variety of activities—such as jogging, bicycling, playing tennis, and so on—can free you from the restraints of physical stress. As you consider your resources, how might you metaphorically let out the cows?

Of course, the body should not be active all the time. About a third of each day, it should be asleep. How many hours of sleep do you get every night? The relationship of sleep to physical stress and to gut distress is so profound, we devote all of chapter 5 to it.

Emotional and Spiritual Stress

Emotional stress can result from internal or external sources, but either way, it can wreak havoc on your digestive health. The topic is so important that we devote all of chapter 11 to this crucial but widely misunderstood topic.

Whether or not you are religious, the potential for spiritual stress cannot be overlooked. Every chronic health concern, certainly including gut distress, is a potential spiritual crisis as well as an opportunity for spiritual growth.

Especially when challenged with a life-disrupting health condition or another form of suffering, people are challenged to ask deep questions. How could this happen to me? Why me? Why now? Who am I? How do I fit into the world? What does fate hold in store for me? What is worth doing? Why should I care? Does God exist? Does God have a plan for me? Is it just me against the Universe? What is the meaning of my suffering?

Spiritual distress comes from challenges to our most important connections: with ourselves, with others, with nature, with a higher power, and with a story that goes far beyond our own.

For those who believe in a personal God, major challenges in life may prompt questions related to that relationship. Why doesn't God listen to me? Am I being punished? Why aren't my prayers enough?

For all spiritual distress, the challenge is to translate deep emotions of brokenness or broken connections into spiritual questions. This can often be done in partnership with a trained professional such as clergy, trained chaplains, and spiritual directors. Well understood by all spiritual care professionals is that the act of giving voice to spiritual questions, concerns, and challenges can itself be healing. Also well understood is that for *your* questions, the best answers are found rather than given. These professionals can help you navigate your path to find your own answers.

Pharmaceutical Stress

Pharmaceutical stress is another name for the visible and invisible side effects of prescribed medications. Yes, medications that have the power to do good also have the power to harm, tax, deplete, or strain the body. Certainly, many medications can cause gastrointestinal distress, rashes, headaches, and the like. But medications can also cause unrecognized nutritional problems that can affect energy, mood, memory, sleep, and general oomph. Ironically, additional medications are often prescribed to treat the side effects of the first medicine.

If you require prescription medications, be on the lookout for side effects. These can magnify if you are taking more than one medicine at a time. Certain medicines are more likely to have an effect on your gut distress issues. Have you required antibiotic medications in the past year? Past five years? Have you been taking medications (such as Prilosec or Prevacid) to reduce reflux disease or gastritis for more than one month? Have you been taking prednisone or other immune suppressing medications within the past year? Are you undergoing chemotherapy as part of cancer treatment, or have you done so in the past three years? Has your doctor assessed you for nutrients potentially depleted by prescription drug use? (See *www.trustyourgutbook. com* for a complete listing to help you prepare to talk with your physician.)

If you suspect that you are suffering from pharmaceutical stress, see your primary care physician or find an integrative physician. These types of problems require a professional who understands the chemical reactions in your body. The physician can look for alternative medicines or other ways to treat your condition while minimizing the stressful side effects. Yoga and deep breathing are not going to do the job in this dimension of stress.

Dietary Stress

Because everything you eat and drink goes right to your gut, your observation of potential dietary stressors is a crucial component of the CORE program. What could be more relevant to your gut distress? Key concerns here are:

1. Potential food allergies and/or food reactivities
2. Nutritional insufficiencies
3. Oxidative stress

Additional concerns include food cravings, mindfulness, and the quality of foods eaten. These will all be addressed in more detail in the next chapter.

To begin your dietary stress assessment, keep a diary of everything you eat for at least two weeks. Also note when you eat, how you eat, and why you eat. How often do you skip meals? How often do you snack? What are your snack foods? How often do you cook? How often do you eat whole foods? Processed foods? How often do you eat out? How often do you eat a meal at your desk? While walking? While driving? These aspects of eating provide insights into how mindful your meals are.

Getting a handle on what we ingest and why is complicated because we eat for many different reasons. Food is necessary for survival, but most people make their food choices out of pleasure or convenience. Consider the context of your eating. Do you eat because of emotional cravings? Are your meals largely social in nature? Or are they in a rush? What is the ratio of cooking time to eating time? Are there any foods that you find don't digest well or otherwise disagree with you but that you eat anyway?

As you observe your diet, you may find yourself taking a more mindful attitude toward eating and making time to sit down and savor your meal. A surprising fact: the more we appreciate

Food is necessary for survival, but most people make their food choices out of pleasure or convenience.

the sensual nature of our meals, the more we seek better quality. This fact makes wise choices easier. And you'll be less likely to eat something that would be embarrassing to write down.

One study found that people who kept food journals ended up losing weight. It seems they avoided terrible foods because they were too embarrassed to write them down! But we want you to be honest about observing what you eat and drink. We promise it will be quite helpful.

It is important to keep in mind that these five areas of stress overlap. As you track down the sources of your stress, it's okay to identify more than one area at the same time. In real life, these stressors gang up on you simultaneously.

Oxidative Stress and Antioxidants

Oxidation is a natural process in which exposure to charged-up oxygen (free radicals) promotes decay. You see oxidation when a cut apple turns brown or a bike left in the rain starts to rust. Oxidation also stresses human tissues unless antioxidants are available to counteract the effects.

Antioxidants are the antidotes to oxidative stress. For example, to prevent a cut apple from turning brown, brush on some lemon juice. The vitamin C in lemon juice works an antioxidant, and the apple stays looking fresh. If you don't have enough antioxidants in your body, oxidative stress can damage your cell membranes or even your DNA.

Antioxidants are found naturally in vegetables (carotenoids/pre-vitamin A), fruits (vitamin C), and nuts (vitamin E). The body makes its own antioxidants, too, including glutathione, taurine, and co-enzyme Q10. Many processed or fried foods as well as heavy metals such as mercury or lead, can overwhelm your natural anti-oxidant defenses. You can self-assess your diet in the next chapter, Observe Your Diet.

Joe Hits the Stress Jackpot

Joe was having a hard time at his job because of long hours and excessive demands from his boss (environmental and emotional stress). His stomach was frequently upset, and he found himself taking Tums on an increasingly frequent basis. Because of the long work hours, he ended up getting five hours of sleep per night instead of the eight hours he needed (physical stress).

After several weeks of this routine, he could not sleep normally. He went to his primary care physician, who prescribed a sleeping pill and offered him a prescription for an antacid medication. Joe declined the latter but kept raising the dosage of his sleeping medication until he was able to sleep through the night. At the higher dosage, he started having side effects of increased constipation and daytime sleepiness (pharmaceutical stress). This made it even harder for him to perform well at work, which led to him feeling increasingly overwhelmed (emotional stress).

His depressed mood increased, and over time he ceased to feel any sense of meaning in his work or his daily life (spiritual stress). Before long, he started to lose his appetite, skip meals, and buy energy drinks and processed foods from the vending machines at work (dietary stress).

You may not have as many stressors converging on you as Joe did, but it is common to have multiple stresses bundled together. Through observation, you can disentangle the stresses in order to address them one by one.

If you feel you need additional guidance in finding and addressing your sources of stress, you may want to take our stress assessment, which you can find on our website at *www.trustyourgutbook.com.*

Depletion Syndrome

Long-term exposure to significant stress—positive or negative—can leave many people in a state of exhaustion. This is a low resilience state in which even minor stressors can exceed the body's capacity to respond. It's like the metaphor of the straw that broke the camel's back. Depletion syndrome is measured in the clinic via cortisol patterns in saliva and blood tests for DHEA-S (dihydroepiandrosterone sulfate) and, in women, morning testosterone.

Diagnosis and treatment should be done with an informed and licensed health care professional. In general, centering activities found in this book, including meditation, prayer, yoga, and mild daily exercise are very helpful for allowing the body to return to balance. Health professionals may prescribe helpful herbal medications such as licorice, ginseng, rhodiola, cordyceps, or Ashwagandha. We also sometimes prescribe DHEA supplementation. However, each of these is not without significant side effects and requires medical diagnosis and monitoring

4

Observe Your Diet

Tell me what you eat and I will tell you who you are.

—Jean Anthelme Brillat-Savarin, 19th-century French writer

What could be more relevant to the well-being of your gut than all the food and drink you put into it? You began the process of observing your diet in the previous chapter in regard to dietary stressors. In this chapter, we will go beyond that basic awareness to correlate what you consume with any physical, emotional, or mental symptoms you're experience.

Various foods and liquids can influence how you feel: body pains and headaches, fatigue and general energy level, moods and emotions, and so on. Your diet, then, could be the cause of chronic intestinal distress or a contributing factor that makes your distress worse.

Finding whether something you eat is the culprit behind your agony requires some detective work. Most foods are not harmful for most people. But any given individual can have negative reactions to foods that don't bother most people. It is another *Which foods were present at the scene of the crime?* mystery why this is so, but fortunately we don't need to solve that problem; we only need to find the foods that are doing the damage.

Gut Distress and Jewelry Reactions

Nickel is a common cause of allergic skin reactions to jewelry such as earrings, necklaces, and bracelets. Direct contact can be associated with eczema-like skin changes including itchy, crusty, and red rashes with watery blisters. Treatment is simple: avoid contact.

Many people are shocked to learn that nickel is also found in many common foods. This means that people who suffer from rashes when in contact with nickel-containing jewelry may also have gut distress from nickel-containing foods.

Which foods? Unfortunately, some very well-liked, healthy foods are rich in nickel:

- Cocoa and chocolate

- Oatmeal and wheat bran

- Some fruits and berries including pineapple and raspberries

- Some nuts including cashews, hazelnuts, and peanuts

Additionally, nickel found in tin cans or cooking utensils can sneak into foods.

For a complete description of nickel sources, please go to our website, *www.trustyourgutbook.com*.

Digestive Aids: Enzymes and Acid Support

Digestive Enzymes

Many gut symptoms—including undigested food in stool, getting full quickly, excessive burping and belching, heartburn, bloating, constipation, and gas—can be signs of insufficient digestive enzymes.

Digestive enzymes include proteases (to digest protein), amylase (to digest carbohydrates), lipases (for fats and oils), and maltases (for some sugars and grains).

If you suspect poor digestion, numerous commercial products are available to help. Simply follow the instructions to take the supplements with meals.

Acid Support

Some digestion products contain stomach-acid support in the form of betaine HCL. This product, alone or in combination with other digestive enzymes, can be very helpful in low stomach-acid conditions. If taken alone, start with one tablet with each meal. If you experience burning or discomfort, drink some baking soda in water; take one less tablet with all future meals.

We need stomach acid to break down food, especially proteins. Stomach acid is also important for absorption of several vitamins and minerals. And it is an important messenger when it comes to organ communication for gall bladder and pancreas function.

Low stomach acid is associated with overgrowth of bacteria in the small intestine. This can result in nausea, bloating, vomiting, and diarrhea. Low stomach acid can come from medications (antacids including H2 blockers and PPIs [proton pump inhibitors]), *Helicobacter pylori* infection, low dietary zinc or niacin, autoimmune disease, or chronic inflammation of the stomach. Several other rare causes exist.

Stomach acid production can be stimulated by the aroma of food cooking. You can also aid in its production by consuming apple cider vinegar or concentrated lemon juice. To use, add one teaspoon to a glass of water before each meal. Expect a bitter taste.

The first step in our investigation is to find out which foods were present at the scene of the crime. To do this, create a diary for two weeks in which you keep track of what you eat and what symptoms you have throughout each day. This is called the food and symptom diary. Fourteen days is enough time to start establishing correlations between certain foods and your symptoms. This doesn't necessarily prove guilt, however. It may be a coincidence that you get a headache and gas every time you eat ice cream. Or it could mean you have a dairy intolerance. To establish a level of proof that will hold up in court, you need to eliminate the suspected offenders from your diet for a couple of weeks and see how you feel. Then reintroduce the food to your diet and check for any symptoms. The effectiveness of this process is amazing. We have countless stories of people who recovered from chronic problems once they removed these foods from their cupboards and put them behind bars.

Another value of the food and symptom diary is that it will show you whether you eat the proper balance of nutritious foods. Deficiencies of various vitamins, minerals, and proteins can have astounding negative effects on your body/mind system. Many people improve the quality of their diet after paying attention to what they eat and writing it down in their diary.

The Food and Symptom Diary

The most important tool for observing diet is the food and symptom diary. In our experience, diaries are incredibly helpful tools for disciplined observation.

Dr. Plotnikoff learned the power of diaries firsthand while he was working on a study of menopausal hot flash management. In this trial, 180 women kept a daily record of the frequency and severity of their hot flashes. The result: significant insight into triggers of their hot flashes. They reported that simply keeping the diary helped them make changes in their lives that significantly reduced their hot flashes. These women were highly motivated and highly educated, yet, before keeping the diary, they did not have the advantage of perspective. They had not made any connections until they wrote it all down. The diary keepers reported a much stronger sense of control over the frequency and intensity of their hot flashes than the non-diary keepers did.

The capacity for such diaries to provide enlightening perspective is evident in the case of Janice.

Janice Finds the Offenders

Janice had suffered for more than fifteen years from fatigue, muscular pains, and dizziness—plus significant gastrointestinal problems. She had frequent, unpredictable but urgent bowel movements as well as bloating significant enough to affect her clothing choices. Gas production was frequent, bothersome, and noticeable. She had plenty of sleep and went through at least twenty prescribed medications. She tried yoga, meditation, and even physical therapy. Nothing seemed to work.

She noted, "I eat a lot of Tums because I feel uncomfortable from heartburn and bloating. In fact, I eat a lot of them because everything is uncomfortable."

When she came to Dr. Plotnikoff, he gave her some homework: a food and symptom diary. On her return visit, she noted that the diary yielded an important new awareness: "I do not feel good very often." The diary also allowed Janice to make important connections. She saw that eating yogurt was routinely followed by headaches, generalized pain, and congestion. Caffeine was associated with soreness. And with cake, she felt "really miserable and everything just hurt."

The diary provided perspective and allowed Janice and Dr. Plotnikoff to generate a testable hypothesis: would Janice benefit from eliminating gluten/wheat and dairy from her diet?

Two months later, Janice had tested the hypothesis and had her results: she saw significant improvement in regard to bloating, bowel movements, and gas when she did not consume gluten or dairy. Additionally, she reported a "really big reduction in pain over the past seven days. In fifteen years, I have never gone seven days with no pain. I have not required a pain med all this week, and I previously used one daily!"

How to Keep a Food and Symptom Diary

Use a notebook to keep track of the following items for two weeks to develop an awareness of any connection between what you eat and how you feel. Your goal is to identify the value of an elimination diet or of formal testing for food allergies or reactivities. Your diary should include five general categories of data:

1. Date
2. Time and detailed description of food and drink
3. Time and presence of symptoms
4. Duration of symptoms
5. Degree of interference with daily activities

What do we mean by symptoms? Use the following list as a guideline, but remember that all symptoms count, no matter how unusual they may seem.

GI symptoms to note:

- Bad breath
- Burping, belching, hiccuping
- Heartburn/reflux
- Appetite
- How quickly you fill up
- How quickly you are hungry again
- Nausea
- Vomiting
- Abdominal bloating
- Abdominal cramping
- Bowel movements: frequency, urgency, strain, consistency, buoyancy. Do they contain any undigested food, mucus, or blood?
- Intestinal gas: frequency, level of bother

- Weight gain
- After eating: likelihood of exercising, ability to function effectively

Additional symptoms for the symptom diary:

- Brain fog: concentration and memory difficulties
- Congestion (nose, sinus, chest)
- Pain (joints and/or muscles)
- Headaches
- Fatigue
- Fluid retention (swelling, edema)
- Mood challenges: mood swings, depressed mood, anxiety
- Sleep challenges

The Usual Suspects

The most common foods for adverse reactivity are:

- Wheat
- Dairy
- Corn
- Nightshades (tomatoes, potatoes, eggplants, and peppers—including spicy and bell peppers)
- Legumes (all kinds of peas and beans, including soybeans, lentils, and peanuts, but not cocoa or coffee beans)

Extra Benefits of the Diary

Food and symptom diaries enable insights into gut distress as well as into many symptoms found outside of the gut. For this reason, keeping a detailed list of symptoms anywhere in the body makes sense. Many patients have been shocked to see reductions in significant symptoms once they quit eating offensive foods—thanks to the food and symptom diary with a follow-up elimination diet.

The food and symptom diary also provides excellent insight into a healthy relationship with food and eating. Sandy is a good example of someone who was too busy to even think about what she ate—until she used a food and symptom diary.

Sandy's Always on the Go

Sandy was a mother of three who was referred to Dr. Plotnikoff because of a persistent cough and GI concerns that included heartburn, significant burping after meals, nausea, bloating, intense abdominal cramping, and frequent trips to the bathroom with explosive diarrhea. Gas was frequent, bothersome, and noticeable. She rated life interference from the abdominal pain as a six or seven on a scale of one to ten. Her heartburn was so severe that for years Sandy took a prescription antacid medication.

She also reported bothersome fatigue and significant sleep disruption with restlessness, awakening, and rumination. She said, "I almost always worry over something that really doesn't matter."

When Dr. Plotnikoff asked her about any known allergies, Sandy reported tree allergies including birch. Because of the birch allergy—which can be linked to food reactivities—and a history of strong dairy reactivity, Dr. Plotnikoff prescribed a food and symptom diary. The diary led her to recognize clearly for the first time a number of factors that were undermining her health and well-being.

"I struggle with mornings," she observed. "I prepare breakfasts for the kids and then don't have time to eat myself." She also noted that she tended to snack a lot, have a big lunch, but then not eat dinner.

Sandy realized she ate too often at fast-food restaurants and did not eat many vegetables. Through her diary, she understood that she was eating too many carbohydrates, especially breads and pastas. She observed clearly that she did not get a balanced diet. "I'm always on the go." Additionally, she noted that within four hours after eating cheese, she experienced explosive diarrhea and bad stomach cramps.

As a result of the food and symptom diary, she became aware of dietary changes necessary for her health and well-being, including more vegetables and less dairy.

We will return to Sandy's story later in this chapter. The lesson for us now is how the food and symptom diary can provide insight into dietary stressors as well as opportunities for moving toward a healthier state of being.

Cross Reactivity

Formal allergy testing is frequently limited to environmental allergens, excluding food allergies. But many common environmental allergies are linked to food reactivity or allergies. Our body can mistake some foods for pollen and react. If you have seasonal allergies, including hay fever or allergy-induced asthma, or if you have documented birch, grass, ragweed, or mugwort allergies, then you may also react to foods on the list below. Pay particular attention to these in your food and symptom diary:

Season	Environmental Allergen	Related Food Allergies
Spring	Birch pollen	Apples, carrots, celery, hazelnuts, peaches, pears, raw potatoes
Summer	Grasses	Tomatoes

Season	Environmental Allergen	Related Food Allergies
Late summer	Ragweed Pollen	Bananas, melons including watermelon, tomatoes
Fall	Mugwort	Apples, broccoli, carrots, chestnuts, celery, hazelnuts, kiwi, peanuts, peppers, sunflower seeds, and some spices including parsley, caraway, cumin, coriander, fennel, and anise
Year-round	Latex	Apples, avocados, bananas, carrots, celery, chestnuts, kiwi, papaya, potatoes, tomatoes

The food and symptom diary provides an excellent means for self-assessment of the quality of your diet.

The following guidelines support constructive changes in diet and lifestyle that we will cover further in parts III and IV. Level One Awareness provides the prompters for the diet of most North Americans. This checklist is the minimal level of assessment that people should have for good health. The goal is awareness. For example, how much food is processed? How many "empty calories" devoid of nutritional value? Optimal diets minimize inflammation and oxidative stress. (See *www.trustyourgutbook.com* for Andrew Weil's anti-inflammatory food pyramid and a discussion of the harms found in the Standard American Diet [SAD].) Level Two Awareness represents the dietary ideals for those who are not vegan or vegetarian. Level III Awareness is for those with specialized diets who need to pay extra attention to ensure adequate nutrition.

Level One Awareness (Assessment of the Standard North American Diet)

Daily servings of:

Vegetables

Fruits

Processed grains: flours, breads, pastas

High fructose corn syrup, concentrated sweeteners, artificial sweeteners

Caffeine

Dairy, by type: butter, milk, cheese, yogurt, cottage cheese

Hydrogenated vegetable oils

Calories from:

Nonalcoholic beverages

Alcoholic beverages

Weekly servings of:

Fast food

Fish

Lean red meat

Level Two Awareness (Self-Assessment of Advanced Diet with No Gluten, No Dairy)

Daily servings of:

Antioxidants: fruits and vegetables with deep colors:

- Blues and purples: all dark berries including blueberries, bilberries, blackberries

- Reds: bell peppers, cherries, cranberries, pomegranate, raspberries
- Orange: autumn squash, bell peppers, carrots, sweet potatoes
- Yellow: bell peppers, zucchini

Cruciferous vegetables: bok choy, broccoli, broccolini, broccoli rabe, brussels sprouts, cauliflower, collards

Non-dairy sources of calcium: almonds, beans, blackstrap molasses, broccoli, dark leafy greens (bok choy, collard greens, kale, mustard greens, turnip greens), dried figs, okra, tahini, tempeh

Traditional grains: buckwheat, quinoa, millet, teff, wild rice

Weekly servings of:

Avocados and olives

Extra virgin olive oil for cold cooking: dressings, dips

Fermented foods: sauerkraut; lacto-fermented pickles, beets, or onions; unpasteurized yogurt, kefir, kombucha, kimchi, miso, natto, tempeh

High oleic acid sunflower oil or coconut oil for hot cooking

Mercury-free fish

Organic, grass-fed, antibiotic-, hormone-, and pesticide-free beef, lamb, poultry

Level Three Assessment (Standard Vegetarian Diet with No Dairy, No Meat)

Daily servings of:
Complete protein: egg, quinoa, or a combination of beans, grains, nuts, and seeds as well as supplements including tahini

Note: Soybeans and legumes are low in amino acids tryptophan and methionine, both of which are crucial for mood, energy, and sleep, so they are not listed in this offering of complete protein choices. Grains, nuts, and seeds are low in lysine, which is crucial for tissue repair, hormone production, and vitamin B_6 function for serotonin production, so supplement them with other forms of protein.

Sufficient protein: Aim for at least 0.35 grams of protein per pound of weight for adults.

B_{12}: nutritional yeast, fortified cereals, vitamins

Long-chain omega-3 fatty acids EPA and DHA: spirulina and other algae for DHA; walnuts and flax seeds along with low-fat diets for EPA production

Iron (for menstruating women or anyone with iron loss): blackstrap molasses, cooked soybeans, lentils, lima beans, quinoa, spinach, Swiss chard, tempeh, tofu

Sandy Stops Coughing

The insights gleaned from Sandy's food and symptom diary provided the groundwork for changes in her diet that could alleviate her GI symptoms and her cough. Despite the dietary changes, including complete elimination of dairy, she still had significant symptoms. "It's not only my gut. I have been struggling for years with this cough," she said. "People think I am a smoker or think I am sick. If it's not my gut, its my cough. They get in the way of everything, including my job."

This cough also made it difficult for Dr. Plotnikoff to do his job. The cough's persistence meant that he and Sandy could not recognize whether there was any relationship between food and symptoms. She was always coughing no matter what she ate!

This left them with two options: a complete elimination diet (no wheat, corn, dairy, legumes, citrus, or nightshades) or specialized laboratory testing for food reactivity. She chose the latter.

Sandy's blood test results were quite amazing. They showed no evidence of any immunoglobulin E (IgE) food allergies. They did, however, demonstrate significant evidence of immunoglobulin G

(IgG) food reactivity to corn. From her previous food and symptom diary, we could see that corn, including corn hidden in lists of ingredients,[1] were a significant part of her daily diet. The next step was clear: no corn until evaluation at the next visit.

Upon her return two months later, Sandy reported a nearly complete resolution of her cough. "First time in many years!"

Sandy also showed marked improvement in her energy levels, and she was sleeping through the night. Her GI distress nearly went away: no burping, no nausea, no vomiting, no abdominal cramping, and no episodes of explosive gas or diarrhea. She had just a little bit of bloating, and her gas became much better—less frequent, not bothersome, and not noticeable.

So, in Sandy's case, observation via the food and symptom diary plus specialized laboratory testing resulted in complete recovery from her years of both gut distress and her persistent cough.

Food Allergy and Food Reactivity Testing

Laboratory testing is available for two forms of immune response. The first form, immunoglobulin E (IgE), is what allergists commonly test. This can be done through blood testing or skin prick testing.

When people have an IgE food allergy, such as a peanut allergy, when they come in contact with the allergen, they may develop life-threatening hives, asthma, or throat tightening and need to use a shot of epinephrine and go to the emergency room.

Remember: Ig*E* allergies are associated with *E*-pinephrine and *E*-mergency rooms.

Allergists primarily serve patients with severe food allergies. They tend to be very conservative in making food allergy diagnoses with less than severe reactions.

If people have an immunoglobulin G (IgG) food reactivity, the manifestations are more delayed, less dramatic, and less clearly

---

1 See *www.trustyourgutbook.com* for a more complete discussion of hidden triggers in food, such as corn, wheat/gluten and MSG.

connected. When people have an IgG food reactivity to peanuts and eat them, they may have nonspecific symptoms such as headaches, weakness, achiness, mood swings, or abdominal discomfort. The food and symptom diary can often help with this.

IgG reactivities are not allergies. They are associated with statements such as, "Gee, I did not realize I was reactive to that food."

Very few board certified allergists use IgG food reactivity testing. Both IgE and IgG food reactivity tests are available through other qualified health professionals.

Three Dietary Supplements for Gut Distress Symptom Relief

These three dietary supplements have been well studied and have been shown to be effective for relieving gut distress. They do not require a prescription; just follow the manufacturer's guidelines for dosage.

- Enteric-coated peppermint-oil capsules: prevention and reduction of abdominal spasms

- Peony-licorice formula: prevention and reduction of abdominal spasms (For intermittent use only: can cause reversible low potassium and high blood pressure with daily use).

- Iberogast (nine-herb product): reduction of abdominal pain, control of gut hypersensitivity, normalized movement of the bowels

Observing your diet is an essential step toward achieving wellness, and it requires ongoing vigilance. We have a relationship with food that is often more emotional than rational. We are driven by hunger, but also by routine, busy schedules, food cravings, and peer pressure to eat things that trigger gut problems. As you routinely observe what you eat, you will find it easier to choose healthier foods and to avoid troublesome foods.

5

Observe Your Sleep

Sleep is the best meditation.

—His Holiness the Dalai Lama

Chronic gut distress has a very close relationship with sleep problems. How well can you sleep if abdominal cramps keep waking you or if diarrhea forces you to get up several times? Likewise, if you have a lousy night of sleep, you're more likely to suffer gut symptoms the next day because you are exhausted and out of sorts to begin with. A vicious cycle can develop because the extra stress of gut problems can hinder sleep, and a poor night's sleep contributes to the body/mind imbalance that promotes gut distress in the first place. The purpose of observing your sleep is to understand your sleep-wake cycles and the many factors that may affect your sleep. The goal is to understand where adjustments may be made to improve your gut distress and your overall health.

Julie Struggles to Sleep

The quality of Julie's sleep started going down when her chronic gut distress began, several years before she came to see us.

She told us, "For the first several months that I had gut problems, the pain bothered me," she said, "but I was still able to sleep reasonably well. But after about six months or so, it's like my whole system became more sensitized, more on edge. I started to feel more discouraged and generally out of balance."

After a while Julie started worrying about her sleep as much as she worried about her gut pain. She entered a frustrating cycle that repeated itself nearly every night. About seven or eight pm, she would start dreading going to bed. She'd worry about whether she'd be able to fall sleep and actually make it through the night. Would she be kept awake with pain or up going to the bathroom?

"When I finally get to bed, I'm so exhausted that I fall asleep within ten minutes. But then, about two hours later, I may roll on my side and a jolt of pain in my side wakes me up. Next thing I know, I'm wide awake." She'd lay there worrying for half an hour before falling asleep again. She'll wake up like that a couple more times. When the alarm goes off, she is so tired and discouraged she can hardly get out of bed.

Sleep Observations

Julie's experience and variations of it are fairly common. In every case, the solution begins by observing your sleep rhythms and patterns as well as your activities and habits in the hours before bedtime. As you become aware of the various factors that affect the quality of your sleep, you will be able to make adjustments to improve your sleep, and thereby help your gut symptoms heal, too.

Your sleep observations fall into three general categories:

1. Your sleep patterns

2. Your emotional and cognitive responses to sleep distress

3. Your sleep hygiene

Sleep Apnea

Sleep apnea is a chronic medical condition where a person repeatedly stops breathing during sleep. These episodes last ten seconds or more and cause oxygen levels in the blood to drop.

Obstructive sleep apnea is caused by obstruction of the upper airway. Sleep apnea is due to the failure of the brain to initiate a breath. If you find that you wake during the night due to snoring or the feeling of difficulty taking in sufficient air, you might have apnea.

Has anyone ever told you that you snore or sound like your breathing is irregular or restricted? If so, see a specialist at a sleep clinic. The CORE program encourages a medical evaluation for any apnea symptoms.

Observe Your Sleep Patterns

The most important thing about sleep patterns is that our body seeks out natural rhythms each day. Are your sleep patterns consistent or erratic? Do you go to bed at the same time every night, or is it drastically different from week to weekend? To get a handle on your personal schedule, it is useful to start keeping a sleep diary so you can chart the timing and flow of your sleep. When do you go to bed and when do you get up? Do you need an alarm clock to wake you, or do you get up on your own? Do you sleep solid, or do you wake up often?

Environment/Sleep Cycle Disconnects

Special sleep cycle challenges can come from working night shifts or changing time zones and experiencing jet lag. These environmental changes result in circadian (twenty-four-hour cycle) rhythm disorders, which significantly increase your risk of gut symptoms including pain, constipation, and diarrhea. As documented in nurses, the connection between rotating night shift work and gut distress is independent of the quality of sleep.

Two ways to help restore normal circadian rhythms are by taking melatonin, a hormone naturally produced by the brain in response to darkness, and getting light therapy, the use of full-spectrum light boxes to simulate sunlight. You can take a dose of 3 milligrams of melatonin several hours before bedtime. This will not cause sleepiness; it only tells the body that it is dark outside. Light therapy, using a phototherapy light box or visor, is done for up to thirty minutes upon waking. Other ways to improve your twenty-four-hour rhythms include exercising after waking up and eating meals on a regular daily schedule

---

Everyone's big question is: how many hours of sleep do I need? We've heard several nationally known experts on sleep repeat a funny truism: "Recent research has determined exactly how much sleep we need—about ten more minutes after the alarm goes off!" That may be about as accurate as we can get, because there is no single answer that works for everybody. A lot of studies suggest around seven and a half to eight hours, but it depends on the person. That means you have to figure out how many hours of sleep work for you. It makes a big difference in your health and your state of mind if you don't get enough sleep.

Do you feel truly rested each morning? Or do you always feel like you need a nap? How many hours of sleep do you think you need? Are you getting that amount? A half hour of extra sleep each night can make an amazing change. It's the difference between groggy and functional. When you are on vacation and you can sleep as long as you'd like, how many hours do you slumber?

Some people know how much sleep they need, but they don't get it because they are too busy meeting others' needs and ignoring their own. They don't feel *entitled* to a full night's sleep.

Mike's Dilemma

Mike was a forty-five-year-old husband and father who knew how many hours he needed to sleep to prevent the onset of IBS

symptoms like bloating, sharp abdominal pain, reflux, and chronic constipation.

"It's like I'm so fragile," said Mike, "If I get eight and a half to nine hours of sleep, then my gut symptoms seem to be better the next day. But if anything interferes with my sleep schedule, I'm in trouble."

The problem was that he was in trouble a lot—if he slept nine hours a night, then he missed appointments. He felt like he was letting down his boss, coworkers, and family. But if he lived up to all of his family and work commitments, then he didn't get enough sleep. It was a no-win situation.

Mike understood the logic of needing more rest, but he felt that he couldn't let others down by taking care of himself. With some gentle prodding from his wife and friends, Mike finally saw a psychologist, who helped him identify why it was hard for him to feel entitled to prioritize self-care. Eventually, Mike learned to protect the eight and a half hours of sleep he needed every night. Within a few weeks, he felt significant improvement in his IBS symptoms.

To promote better sleep and consistent sleep rhythms, it is very important to go to bed at the same time every night and to set your alarm to awaken you at the same time every morning. To figure out when to go to bed, listen to your body. When do you start feeling drowsy?

Another factor that affects how many hours of sleep you need is whether you wake up at night. Is it hard to fall asleep when you go to bed? Do you wake up and then find it hard to go back asleep? What wakes you up? Gut pain? Running to the restroom? Can you fall back to sleep easily? Keep track of your good nights and your bad nights, and look for patterns. What can you do to achieve more regularity in your sleep?

What can you do to achieve more regularity in your sleep?

Insomnia

Do you fall into one of these categories of insomnia?

- Initial insomnia: Difficulty falling asleep. After turning off the light, people with initial insomnia may lie in bed for thirty to sixty minutes before falling asleep.

- Middle insomnia: Waking up in the middle of the night. People with middle insomnia fall asleep okay, but then they awaken two to three hours later. They may stay awake from a few minutes to an hour. This could happen up to four times a night. They finally fall asleep until their alarm goes off.

- Terminal insomnia: Waking too early. Terminal insomniacs wake up during the night and then cannot fall back to sleep. They stay awake for the rest of the night until the alarm goes off.

Observe Your Emotional and Cognitive Responses to Sleep Problems

One of the most difficult aspects of sleep disturbance is that the more we worry about falling asleep, the harder it is to fall asleep! More generally, the more we worry about anything, the harder it is to fall asleep. The Dalai Lama identifies sleep with meditation, and that has a lot of truth to it. When you meditate, you clear your mind and try to avoid any thoughts or worries. The same applies to sleep—it comes easiest when you blank your mind.

By taking note of the thoughts and feelings that dominate your mind when you should be sleeping, you can start clearing away the worries to get more rest. What goes through your mind when you can't sleep? Are you worried about sleeping? Are you worried about your gut distress and how it will affect you tomorrow? When you go to bed, do you worry about tasks that didn't get done that day or events that upset you? Do certain scenarios keep replaying in your mind? Do you rehearse what you should have said to someone or what you need to say tomorrow? Do you get angry because you can't sleep? How about sad, depressed, or ashamed?

If your observations tell you that you bring too many problems to bed, you can deal with them earlier in the day so you can relax when you go to bed. Perhaps you can set aside a worry time—a couple hours before bed to give your full mind to your problems du jour. You can try writing everything down to get it out of your system. Start a journal. Make a to-do list for tomorrow. You can try to resolve issues when they happen rather than just absorbing them as if they will go away on their own. (We will discuss additional techniques for soothing difficult thoughts and emotions in Chapter 11.) You can make a deliberate effort to go to bed with a clear mind. Even if your main worries are your gut distress, you will feel much better just by knowing you are starting to do something about it.

Janine the Early Riser

Janine was a corporate attorney in her early fifties who was under intense pressure to put in as many billable hours as she could. She worked from early morning to late night. "I feel like I just can't afford to be held back by this awful pain and diarrhea," she said. "I have to keep working no matter what. That constant pain in my mid-abdomen feels like such a burden to deal with all day long. I probably have to go to the bathroom at least seven or eight times a day, yet I'm always being scrutinized for how my productivity numbers are measuring up."

With her exhausting schedule, Janine was ready to conk out the moment her head hit the pillow, and she needed about eight hours of sleep to feel rested. However, she rarely got it. Typically, she woke up suddenly with her heart racing after five or six hours. Once awake, she started worrying about the day at work and about how much her gut would interfere with her productivity.

"I asked my doctor to prescribe sleeping medications, but they only helped for a while, and now I'm back to waking two hours early again," she said. Once her worries kicked in, her lower gut pain got worse and there's no getting back to sleep. She lay there in bed, tense and frustrated, until it was time to wake up.

Calm Your Mind and Body with 4-4-8 Breathing

Techniques that reduce anxiety, arousal, and dysregulation of your nervous system can help you sleep better. The 4-4-8 Breathing technique is a simple tool you can use before you go to bed or when you wake up during the night:

1. Settle into a comfortable position, either sitting in bed propped up with pillows or lying comfortably in bed.

2. Close your eyes.

3. Breathe in through your nose and count to four.

4. Hold your breath for four counts.

5. Exhale through your mouth for eight counts.

6. Repeat the cycle of 4-4-8 at least 10 times.

While you do this, keep your attention focused on the counting (as you inhale, mentally count in a rhythmic pattern). If your mind gets distracted during this exercise, bring it back to the counting.

This brief technique will calm your body and quiet your mind enough to rest a little more comfortably. If the 4-4-8 is too difficult for your breathing, or if it brings on a sense of discomfort, shorten it to 3-3-6. We will discuss additional techniques for calming your mind in Chapter 10.

The Thirty-Minute Rule: Don't Try to Sleep

If you awaken during the night, allow yourself thirty minutes to stay in your bed and practice relaxation or breathing techniques. If you are still awake after thirty minutes, you should not *try* to fall back asleep.

Instead of staying in bed, go to another room where you have a comfortable chair or couch, low lighting, and perhaps a blanket. Get as comfortable as you can and do something that occupies your attention in a gentle way. Listen to soft music or do some light reading. Stay away from newspapers, the TV, or other activities that overly stimulate your mind. Don't *try* to relax, just keep yourself gently occupied. Before long, you will find yourself becoming drowsy again, and when you do, go back to your bed.

Observe Your Sleep Hygiene

Sleep hygiene is the practice of developing healthier sleep habits. It involves making mindful changes in behavioral and environmental factors that surround sleep and may interfere with it. Observe your behavior for the last few hours before you go to bed, and also take note of the quality of your bedroom environment. Awareness of these areas will let you know what needs to improve to allow you to sleep better, which in turn takes unneeded pressure off your gut.

Watch Your Behaviors

The general rule is to avoid stimulating activities before bedtime. Do you drink coffee or some other caffeinated beverage after noon? Keep in mind that caffeine can affect your body for up to twelve hours. Better to stay away from such beverages, and that includes certain soft drinks as well as energy drinks. How about alcohol? The hallowed tradition of a nightcap or a little wine before bed has been refuted by modern science. Alcohol is now well understood to be a sleep disruptor. You may get drowsy and fall asleep, but you don't necessarily stay asleep all night. If you have to drink something in the evenings, try non-caffeinated herb tea. Even then, don't drink it right before bed or you'll wake up in the middle of the night to run to the bathroom.

Eating is another area to keep your eye on. Do you eat dinner very late? Do you eat in bed? Did you know that high amounts of protein tend to stimulate your system? It's best to max out on protein during lunch. Do you know it's difficult to have a sound sleep while your body is busy digesting food? Aim to finish eating and drinking at least a couple hours before it's time to go to sleep.

Smokers, take note. In case you need another reason to quit smoking, here it is: tobacco interferes with sleep. Nicotine is a stimulant, and using it too close to bedtime may make it difficult for you to fall asleep. Smoking also is associated with a disruption of the basic structure of

Smoking also is associated with a disruption of the basic structure of sleep.

sleep called sleep architecture. One research study showed that smokers take slightly longer to fall asleep, sleep less, and have less deep sleep than nonsmokers do. Still want to light up?

Other activities that can overstimulate your body when it is close to bedtime include exercise, whether it's biking, walking, or using a stair climber or elliptical machine. Such aerobic exercise can be very helpful for reducing pain, improving gut motility, and helping with sleep. However, you must be mindful about *when* you exercise. For many people, exercise has an effect that is initially stimulating before relaxing. Therefore, be sure that you end your exercise at least three hours before bedtime. This will help you fall asleep and stay asleep more easily.

Do you keep busy with various activities right up until the moment you go to bed? If you are working on projects, playing video games, watching TV, or surfing the web, your body/mind has been stimulated, and you are not ready for bed. Sometimes the medium is the message—regardless of the content you are watching. A recent study showed that the bright light of a computer screen may alter the body's biological clock and suppress the natural production of melatonin that's critical to the normal sleep-wake cycle as well as other daytime/nighttime cycles. Melatonin is a hormone in the body that helps regulate a person's sleeping and waking hours. This hormone tells the body when it is dark. Researchers say exposure to more light reduces the amount of melatonin produced, and a decline in melatonin production is often blamed for sleep problems. If you do any computer-related work or play in the evening, make sure to stop at least one hour before going to bed.

What do you do to wind down before sleep to get your mind in the right mood? Do you take a warm bath or shower, brush your teeth, or wash your face? If you consciously clear your schedule so that you're not eating, drinking, exercising, or watching a movie, then you have time to develop a bedtime ritual to prepare for sleep. Perhaps you could listen to some soothing music, do some light stretching, turn down your bed, read a book, or do some slow deep breathing.

Melatonin and Gut Distress

Our body's master clock resides in the hypothalamus of the brain. This is the central coordinator of the body's twenty-four hour circadian rhythms. Additional clocks exist in the gut, muscle, and fat cells. None of these run on a perfect twenty-four hour schedule. The body resets these clocks every day through exposure to the day/night light cycle.

Light is the most powerful regulator of the body's master clock, and the hormone that conveys this message is melatonin. Its daily rhythms resynchronize all of the body's clocks.

Disruption of circadian rhythms in the short term can result in fatigue, insomnia, and disorientation. In the long term, disruption is associated with accelerated aging, risk of cancer, and gut distress.

The second brain, in the gut, has several twenty-four-hour biological rhythms. These include gut movements, stomach acid production, production of mucosal barrier protectors, digestive enzyme production, nutrient transport, and immune function.

Circadian rhythm disruption is linked to many forms of gut distress: GERD, indigestion, ulcers, inflammatory bowel disease, and irritable bowel syndrome.

Melatonin supplementation, 3 milligrams by mouth at seven or eight pm, has been shown in clinical studies to restore normal gut motility as well as to reduce abdominal pain, bloating, and cramping with defecation. Melatonin supplementation has been shown in animal models to prevent as well as heal stomach ulcers.

Melatonin production is enhanced by a dark bedroom. By contrast, melatonin production is suppressed for several hours by exposure to light from TV, computer, or smartphone screens. For some, even the little light on the corner of an electronic device in a bedroom can disrupt melatonin production. For the best sleep, keep a dark bedroom, and supplement your melatonin if needed.

Three Seed Tea

Ayurvedic medicine specialist Marcia Meredith, NP, recommends this balancing and nourishing non-caffeinated tea for people who enjoy warm drinks or flavored water.

This is considered tridoshic, or appropriate for all types of people. In Ayurvedic tradition, this tea is used for calming both the mind and the gastrointestinal tract.

Recipe:

Take equal amounts of the uncrushed seeds of cumin, coriander, and fennel. Mix well in an airtight storage container. Pour some of the mix into a tea ball. Steep in hot (not boiling) water for five minutes. As the water cools, add lemon or honey for the taste you enjoy most.

Option: Consider adding anise seed or other flavoring.

Note: Persons with late summer or fall hay fever allergies may have cross reactivity with these herbs.

Herbs and Supplements for Sleep

Numerous herbs have been shown to be safe and helpful for achieving sleep. These herbs help induce relaxation and drowsiness:

- Valerian (*Valeriana officinalis*)
- Hops (*Humulus lupulus*)
- Skullcap (*Scutellaria lateriflora*)
- Chamomile (*Matricaria recutita*)
- Passion flower (*Passiflora incarnata*)

Multiple commercial products contain one or more of these herbs. Chamomile, if used as a tea, should be steeped for at least five minutes to release the active ingredients.

Dietary supplements for the calming of restless minds include taurine (500–1,000 milligrams per day); L-theanine, a derivative of tea (100–200 milligrams a day), and inositol (50–100 milligrams a day).

Mineral supplements for sleep include magnesium (360–600 milligrams a day of citrate, glycinate, taurate, or malate but not oxide) and calcium citrate (500 milligrams a day).

As with all supplements, we recommend consultation with an informed health professional to guide your selections.

Attend to Your Sleeping Environment

Observe your bedroom as the center of your sleeping environment and look for any distractions that may interrupt your sleep. Do you spend a lot of time in your bedroom doing activities other than sleep or sex? Do you watch a lot of television there? Read entire books? Use your laptop to catch up on work or for social networking? This is bad sleep hygiene, and it interferes with a good night's sleep. It is best to take work materials, computers, and televisions out of the sleeping environment. Use your bed only for sleep and sex to strengthen the association between bed and sleep.

Are your pets allowed in your bedroom? Does your cat walk across your head at three am as a subtle reminder that she's hungry? Does your dog sleep at the foot of the bed and sometimes cut off the circulation in your legs? We know you love your animals, but do yourself a favor and keep them out of your bedroom.

Is your bedroom dark enough? Do the shades and curtains block out streetlights and automobile headlights? Is the nightlight too bright? Do you have so many electronic devices around that the little red and green lights make the room look like an airplane cockpit? As we mentioned before, even a small amount of light can disrupt your melatonin production and interfere with your sleep. Are the lighted numbers on your alarm clock staring at you? If you have trouble sleeping, seeing what time it is can add to the worry. Try turning it away from you. Do you need eye shades?

What about other conditions? Is it warm enough? Too warm? Some studies show that insomniacs have a warmer than average body temperature before bed, so it's best to keep your bedroom cool—from 60 to 68 degrees Fahrenheit. Temperatures in this range may help facilitate the decrease in core body temperature that in turn initiates sleepiness. Also, is it noisy? Can you hear noise from outside in the neighborhood? If external noises (or internal noises like snoring or animal sounds) become disruptive, you may want to try some cellulose earplugs. Other environmental problems that may affect your peace of mind include invasive electrical fields generated by electrical appliances or equipment near your bed. Do your best to make your bedroom a haven for sleep and not a torture chamber.

Structure Your Daytime

If you think in terms of a twenty-four-hour cycle, nighttime and daytime are the yin and yang of your life. If you have a hectic day, how can you expect to have an orderly and punctual sleep time? Try to establish a daily activity routine. This sense of order will help to keep you centered during the day so you bring less stress to bed with you at night. Also, just as it is helpful to have a dark room at night, try to expose yourself to as much light as possible during the day, especially in the morning. This slight stimulating effect keeps you awake and can enhance your mood.

Do you enjoy taking a nap every day? Or do you desperately need one? If your sleep is generally good, an occasional nap is fine and won't adversely affect your sleep health. But if you have trouble sleeping at night, routinely napping will only amplify problems. Napping reduces your chances of sleeping a full night. A long nap or a nap taken too late in the day may adversely affect the length and quality of nighttime sleep. For people with sleep issues caused by chronic digestive distress, napping during the day can perpetuate bad sleep habits, confuse your internal clock, and send your insomnia into a chronic spiral. It's better to avoid naps and find other ways to spend your time until bedtime.

Need Professional Help to Sleep?

When your sleep disturbance is so severe that it's difficult to get through the day, you may need to consider additional professional resources. The services of a psychiatrist, clinical psychologist, or other mental health professional may be very helpful to get through a difficult stretch. Prudent, temporary use of certain medications (such as sleep medications or antidepressant medications) may help calm and regulate your nervous system long enough for you to start to get some badly needed rest. Structured psychotherapy may provide the necessary skilled help for you to identify and resolve emotional conflicts that are monopolizing your sleep time. This will give you the opportunity to calm your emotions and your nervous system enough to begin using these other CORE strategies for longer-lasting relief and improved sleep.

The strategies in this chapter for helping you improve the quality of your sleep are techniques you can practice every day and every night. It may be difficult at first to change habits that have been built over months or years, but each morning that you wake up feeling refreshed strongly reinforces this lifestyle change. Supplements such as prescriptions are not for long-term use; they are bridges that allow you sufficient sleep for the strength and endurance needed for success in all the various components of the CORE program.

6

Observe Your Life:
Your Health and Wellness
Autobiography

When your head says one thing and your whole life says another,
your head always loses.

—Humphrey Bogart in *Key Largo*

Everything is connected, and the more you become aware of the connections that affect your health, the sooner you can achieve a proper balance for self-healing. You have learned how quietly observing your gut has allowed you to listen to its cries for attention, without exaggerating its message through self-obsessed worry or physical overreactions. You have observed how various stresses on your life can throw your body/mind system off balance. By correlating your diet with your symptoms, you have opened your eyes to some of the possible causes of the problems in your gut. The examination of your sleep habits has shed light on a major factor that determines how you feel and function during the day. Now you are ready to take a step back, integrate all this, and look at your life as a whole, a process that will illustrate how everything fits together. It is time to start connecting the dots by creating your own health and wellness autobiography.

To truly make progress on the path of self-empowerment, it is crucial to take command of your own story. You are the lead actor in the drama of your life . . . only you are not acting—this is for real. You can't just take a script written by someone else and pretend that is your life. You can try, and perhaps you can do it very well. But you are not out to win an Oscar. Your goal is to harmonize your gut so it will stop ruining your life. That means you can't just sit and passively listen to what your doctor tells you about yourself, because the doctor doesn't know your whole story—unless you map it out first.

It is a sad fact of our current medical establishment that doctors don't always have time to listen because of pressures on them to see several patients per hour. One study showed that patients who start telling their story of what's wrong with them get cut off by the doctor after about nineteen seconds. Shakespeare couldn't get through a single sonnet in nineteen seconds, much less a full-blown soliloquy. One of the greatest authors in the world is no match for a busy doctor. Your physician is not Barbara Walters conducting a two-hour tell-all interview, and asking evocative questions like, "If you were a tree, what kind of tree would you be?" No, the more common role model for most doctors is Sergeant Joe Friday from the old cop show *Dragnet*: "Just the facts, ma'am."

One problem is that there are a million facts, and you are not even sure what all of them are. So the physician narrows it down for you in a tidy review of physical symptoms that takes your medical history very quickly: "Do you have a fever? Chills? Sweat? Nausea? Vomiting? Diarrhea? Constipation? Headache? Discomfort?" This checklist approach provides the efficiency demanded by the system, but as you have found out time and time again—the ultimate conclusion is "We can't find anything wrong with you. You'll just have to adjust to your chronic condition. Here, take a pill and try to relax."

We have learned from working with many patients with chronic conditions that considering just the facts of a person's life doesn't tell us enough. We also need to learn about a person's beliefs and feelings. We have found that only a broad array of autobiographical information can provide enough context to help us figure out what the facts *mean*. It is no wonder that conventional doctors can't figure out what's ailing you: they don't have the whole story. That's just the way the standard health care system works. Doctors are looking for things they know how to fix, but if they can't fix it and if it won't kill you, you fall through the cracks.

Because your story cannot get written in the examining room, you're going to have to write it yourself at home. And unless you know your story, you can't put your life into perspective—a necessary step for self-healing.

Don't worry—you don't need to write a full-length memoir in order to heal yourself. The purpose of your medical and wellness autobiography is to map out the various events in your life— both good and bad, high points and low points—and see how all the dots connect. This can be achieved graphically by using timelines and a body chart. This visual history will show you patterns and rhythms of your life that you have never noticed before. This completely new approach to health history results in a tool you can use to help yourself, but it is also a concise medical history that you can share with your doctor.

For Kevin, Seeing Is Believing

Kevin had suffered from IBS for years, but he was reluctant to admit that his emotions might be connected to his physical symptoms, even though Dr. Weisberg had discussed this with him time and time again. But Kevin's attitude changed when Dr. Plotnikoff had him create personal history and medical history timelines. This visual representation of his life made a difference for Kevin.

"Once I wrote everything down, I could clearly see that my severe sinus problems as a child corresponded to a time when

my parents had intense arguments," said Kevin. "They fought so much that I tried to shut my mind and my ears to all the yelling and shouting, and I repressed my emotions."

Kevin also discovered that his IBS symptoms began about the time he and his wife had marital problems. "When I saw it on paper in black and white, I finally accepted that my stress and anxiety had an impact on my physical symptoms and my gut," said Kevin. "Now I realize that I don't learn well when people tell me something—I have become more of a visual learner. It came from trying not to listen to my parents fight."

Once he became convinced of the link between his emotions and his symptoms, Kevin felt more empowered to deal with his IBS by using the integrated approach of the CORE program. The creation of a visual representation of his life allowed him to finally open his eyes.

How to Create Your Medical Autobiography

Your medical history can be visually represented using timelines and charts[2] to depict two broad perspectives: the *patterns* of your life and the *rhythms* of your life.

The Patterns of Your Life

The visual history to depict the patterns of your life has three elements:

1. A timeline of your personal history

2. A timeline of your health history

3. A body chart of your symptoms

---

2 These were developed in part by a group of very creative students at the Minneapolis College of Art and Design—Jenna Ballinger, Anthony Konigbagbe, John Kozak, Jenny Kunstel, Megan Leitschuh—in conjunction with colleagues at the Penny George Institute for Health and Healing. Their youthful inexperience with the health care system allowed them somewhat of a blank slate, and their creative energies allowed them to think outside of existing boxes.

Personal History Timeline

1. Draw a horizontal timeline on a wide sheet of paper, or use the template at *www.trustyourgutbook.com* to make an electronic version of your timeline. The beginning of the line on the left is your birth, and the far right endpoint of the line is today. Along this line, write down the major nonmedical events and milestones of your life, both positive and negative. Key life events include all experiences that helped you develop or shaped you into the person you are today. If someone was to write your biography, what key events would you most want them to include? Standard events may be school highlights, scouting, graduations, marriages, jobs, or sports. Key events also include any significant traumas or key challenges even if they may not be affecting you emotionally or physically today.

2. Make a shorter vertical line that goes up and down at the beginning of the timeline, or use the template at *www .trustyourgutbook.com* to make an electronic version of your timeline. This is a wellness thermometer that indicates your degree of well-being at various points along your timeline. The top represents your most positive perceived states of wellness, and the bottom is your most negative state of wellness. Place a dot correlating to each event—higher for the good, lower for the bad—to show the relative wellness associated with it.

Your basic timeline will look like this:

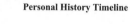

Personal History Timeline

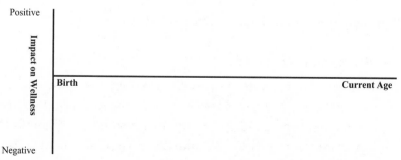

Chronology of life events, milestones, highlights, traumas, etc.

By correlating your wellness to the various events of your life, useful connections or patterns may emerge.

Health History Timeline

1. Create a similar graph, or use the template at *www.trustyourgutbook.com* to make an electronic version of your timeline. Once again, the wellness thermometer is the vertical line, and the horizontal line is your chronology of events, only this time write down your key health events, positive or negative. Standard items include your birth, significant illnesses, surgeries, hospitalizations, or medical workups. Include events related to your greatest strengths for health and wellness as well as events that represent health and wellness barriers you have had or are currently experiencing.

2. As in the previous timeline, above or below each medical or health event, put a dot that represents the height or depth of your perceived degree of wellness at that time.

Personal Health Timeline

Chronology of medical and health-related events.

Bodily Symptoms Chart

1. On the appropriate figure of a man or a woman on the chart below, draw a star on the site of any significant bodily symptoms you experience. You can also note any mental, emotional, or spiritual symptoms on your body chart. Use different colored stars to represent the type of symptoms you experience (red could be sharp, green could be dull, and

other colors could represent pressure, fluctuating pains, burning, stabbing, electrical, and so on). Make larger or smaller stars to indicate the significance or severity of each symptom. Symptoms include any sensations that affect the quality of your health or the quality of your life.

2. Finally, number each symptom. Then on the corresponding vertical timeline, write the number for each symptom at the time when it first developed and also at the time when it changed significantly, either positively or negatively.

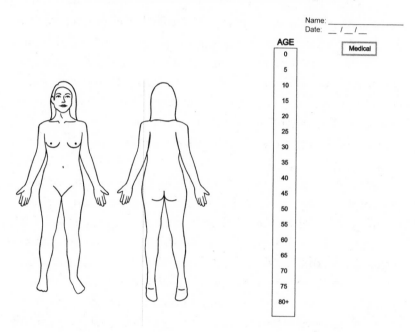

The Rhythms of Your Life

There are five basic rhythms that affect the human condition:

1. Multiple times a day (Ultradian)

2. Daily (Circadian)

3. Weekly

4. Monthly

5. Seasonal

Creating visual representations of your rhythms gives you further information about yourself to help you tell your story and take control of your health. All life operates in rhythms. Some are natural, such as the daily rhythm of night and day, the change of seasons, and the phases of the moon. Some cycles are human inventions that our body/mind has become accustomed to, such as the weekly format of workdays and weekends, and calendar months that vary from the lunar cycle. Some rhythms combine nature and artifice, such as our habit of eating three meals a day. It's natural to get hungry and thirsty, but the three meals a day habit is an invention that has only become standardized within the last couple of hundred years, and the times and sizes of meals vary considerably from culture to culture. Nevertheless, to understand your body/mind and its balance or imbalance, you will need to observe your ultradian, daily, weekly, monthly, and seasonal variability.

Much of the progress of civilization has allowed people to shield themselves from nature—sometimes thwarting your awareness of your natural rhythms. When you live year-round in controlled temperature rooms at 72° Fahrenheit and have ready access to simulated daylight around the clock, or if you take oral contraceptives, you need to spend extra effort to recognize your normal biological rhythms. Ask yourself: To what extent am I sensitized to nature's rhythms? Consider: Do I awaken at sunrise? Do I notice the sunset each day? Do I observe a Sabbath, or do my weekends differ in some way from my week? Have I made time to look at the moon and note its phase each night? Do I plant a garden in the spring and harvest it in the fall? Do my foods match the season?

For each of the five cycles of your life, make a graph similar to the personal history graph. The difference is that the horizontal line measures a different length of time for each graph. When they are combined, these graphs will enable you to learn more about your personal rhythms and their effect on your body/mind. For each graph, note your degree of natural awareness and wellness with a dot above each event—higher for positive wellness, lower for negative wellness.

You've Got Rhythm

Some examples of the rhythms of your life. What additional rhythms would you add to your list?

Ultradian (many times a day)

- Hunger — Meals and snacks
- Urination and bowel movements
- 90-minute energy cycle
- Dream cycle during sleep

Circadian (daily)

- Waking
- Sleeping
- Going to work
- Exercise program
- Hormone release
- Body temperature
- Lung function
- Religious observances and activities

Weekly

- Weekday activities
- Weekend activities
- Sports and recreation
- Religious observances and activities
- Housekeeping chores

Monthly

- Menstruation
- Paying bills and mortgage
- Mood swings

Season

- Holidays
- Religious observances and activities
- Allergies
- Outdoor activities
- Travel/vacations

Ultradian Rhythms

Life undergoes several types of ultradian rhythms throughout a twenty-four-hour cycle. For example, at night, your cycle of sleep stages follows a specific ultradian pattern—including a rotation of light sleep, deep sleep, and dreaming. Because you can detect those stages only with scientific equipment, we are not asking you to graph those. But the daytime rhythm is a different matter. When we are awake, our bodies experiences energy cycles that last anywhere from ninety minutes to two hours, meaning the body oscillates from high-energy to low-energy states repeatedly throughout the day. We also get hungry and thirsty repeatedly throughout a day, and so meals and snack-times should be noted on your ultradian chart.

The tell-tale signs of this cycle include everything from emotional clues like irritability, impatience, and anxiety to physical behaviors such as restlessness, yawning, hunger pangs, urination, and bowel activity. Going relatively unnoticed are changes in heart rate and variations in hormone release. However, when we are under chronic stress, we get overloaded with hormones such as cortisol—which disrupt the effect of the ultradian hormone release cycle and throw your body/mind equilibrium out of whack. By charting the ultradian elements, you can discover whether your rhythms need adjustment.

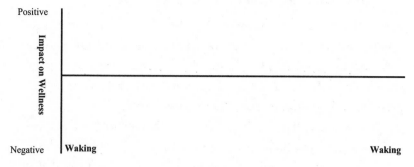

Ultradian Rhythm Timeline

Daily chronology of fatigue, yawning, urination, meals, etc.

Daily Rhythms

The timing of symptoms throughout a twenty-four-hour cycle can provide very helpful clues to your well-being. For this timeline, you need to chart your average times—or range of times—for going to bed, waking up, and any other daily habits. From this chart, you can learn what times of day are generally your best or worst and how much time passes between eating and the appearance of physical symptoms. This timeline also sheds more light on your sleeping habits; be sure to include events such as waking up in the middle of the night. Nocturnal habits are very important for practitioners of traditional Asian medicine, who would recommend looking at unresolved emotional concerns such as grief as the cause for awakening regularly between three and five am.

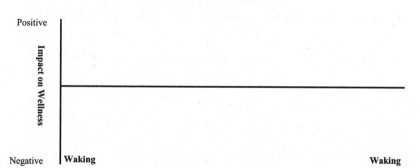

Circadian Rhythm Timeline

Twenty-four-hour chronology of symptoms, daily habits, work, sleep, and so on.

The Chinese Circadian Clock

As far back as the Han Dynasty (202 BC–220 AD), Chinese healers noted that signs, symptoms, and bodily reactions varied throughout a 24 hour cycle. By the 13ᵗʰ century, Chinese practitioners had refined such observations into a 24 hour clock in which each organ's activities are associated with a two-hour period of time.

For example, the lung is associated with 3 am to 5 am. Surprisingly, physicians, physiologists, and patients all confirm that this is the worst time of the day for asthma. The liver, the organ that is believed to direct the flow of *chi* energy, is closely associated with both digestion and emotions, ranging from depression to restlessness. The liver is associated with 1 am to 3 am. This is a common time for people to awaken with emotion or stress-related insomnia.

For the gut, according to Traditional Chinese Medicine (TCM), the large intestine is most active from 5 am to 7 am. This is in preparation for evacuation and cleansing. The stomach is most active from 7 am to 9 am. This is the ideal time for a glass of warm water and a good breakfast. The spleen is most active from 9 am to 11 am. This is the organ that, according to traditional theory, is responsible for digestion and assimilation of nutrients. This is the ideal time for brain activity such as study or work. However, according to traditional beliefs, optimal function at this time requires a good breakfast. And the small intestine is most active from 1–3 pm. This organ also plays a key role in absorption and assimilation of nutrients. This can be supported with the drinking of lukewarm to warm water.

Charting of symptoms and bodily reactions may be helpful for identifying new approaches for a condition. Both TCM and Ayurvedic practitioners are skilled at diagnosis and treatment based in part on evaluation of a person's experience of their 24 hour cycle.

Weekly Rhythms

The seven-day cycle has become ingrained in our culture, and our body/mind has adjusted accordingly. For this chart, try to note your three-month average of activities and perceived degree of wellness for each day. Weekends, start of the work week, mid-week, and end of the workweek represent four periods of time that may vary significantly in terms of your wellness. How and why might you experience these differently?

Weekly Rhythm Timeline

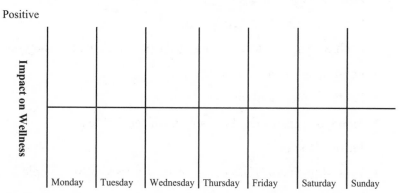

Weekly chronology of work, home activities, recreation, etc.

Monthly Rhythms

Monthly variations in wellness may relate to finances, paydays, deadlines, social events, and so on. When are you usually at your best? Is there a pattern to when you are not feeling your best? Of course, monthly rhythms have an added biological cycle for menstruating women, who ought to chart menstrual timing, regularity, flow, clotting, and cramping—all of which are significant clues to health and well-being. Women with gastrointestinal distress or IBS often report that their symptoms worsen with menstruation. Is this true for you? Women typically report various degrees and different types of uterine pain (sharp, throbbing, dull, nauseating, burning, or shooting) that precede or accompany menstruation. Likewise, women often report symptoms consistent with premenstrual syndrome (PMS) and premenstrual dysphoric disorder (PMDD).

A wide range of physical, emotional, and psychological events should be considered on your monthly timeline. Be sure to note any variations in intensity, impact, or duration throughout the month:

- Bloating, breast tenderness, headaches, and joint or muscle pain
- Disinterest in daily activities and relationships
- Fatigue or low energy
- Food cravings or binge eating
- Feeling of sadness or hopelessness, possible suicidal thoughts
- Feelings of tension or anxiety
- Feeling out of control
- Mood swings marked by periods of tearfulness
- Panic attacks
- Persistent irritability or anger that affects other people
- Problems sleeping
- Trouble concentrating

Monthly Rhythm Timeline

Positive

Impact on Wellness

Week 1   Week 2   Week 3   Week 4

Negative

Men: Use either a calendar month or a lunar month as the timeline to chart your events.

Women: Use the length of your menstrual cycle as your timeline.

Seasonal Rhythms

Seasonal illnesses are very common. For example, winter is the most common time for influenza and seasonal affective disorder (SAD). Pollens emerge in the spring and fall. Are there seasons in which you do particularly well or not well? Do they correlate with time spent outside or inside sealed spaces? For the seasonal timeline, chart your three-year average of activities and perceived sense of wellness.

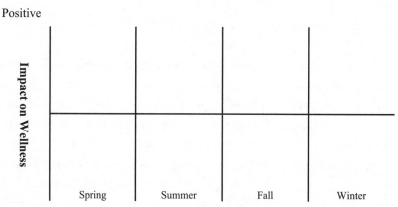

Seasonal Rhythm Timeline

As you create these charts and timelines, you're laying your life out in front of you, and it will be easier to observe yourself objectively. It will become obvious that your life is indeed made up of patterns and cycles. Some connections will jump out at you, as they did for Kevin. Others may be harder to see. Contemplate your charts and compare different timelines to see what items may coincide or overlap. Then you can start finding clues of behaviors, habits, events, or trends that affect your well-being, and you can start thinking about where you can make changes to improve your health.

For example, if you notice that you get hay fever every spring, perhaps you have birch tree allergies. This means that you are likely affected by related food reactivities that could be affecting your gut. This should be checked out.

If your ultradian chart shows you spend a solid six hours at your desk every day without getting up for breaks, then you are defying your ultradian rhythms and are working at low energy levels. One recent study featured in the *Harvard Business Review* suggests that you are better off managing your energy levels than your time, because people who had better energy levels in the study performed more efficiently and more productively than those who worked longer hours with less energy.

Likewise, if you often wake up between one am and three am, remember that this will affect your job performance. It sounds obvious, but we have seen many patients who have learned to ignore the obvious as a way to cope with chronic pain and discomfort.

As you look at your weekly timeline, notice your best days and worst days of the week. Are Mondays bad? Are weekends wonderful? How do you explain your good days and bad days? Monthly cycles are especially important for menstruating women, who should try to pinpoint where on their cycle that flare-ups of distress occur the most. Monthly cycles also coincide with financial events or obligations—such as when the mortgage is due, when paychecks arrive, and the day to pay all your bills. Are economic pressures affecting your cycle of monthly symptoms?

Seasonal events that can affect your mood and distress levels include the holiday season from Thanksgiving to New Year, with all the parties, the opportunities to overindulge in food and drink, family get-togethers, not to mention the pressure of shopping for presents. What's the best season for your mood and energy level? When do you feel most alive—when it's hot or when it's cold? It's surprising to us how many patients report extra pressures on the anniversary of the death of a loved one or some other tragedy. Such annual events can bother people emotionally for years.

Now that you've opened your eyes, keep them open.

As we come to the end of the section on observing, there are two things to keep in mind. First, now that you have taken a look at the various facets of your life, including historical patterns and cycles, it doesn't mean you are done. Now that you've opened your eyes, keep them open. Nothing stays the same in this ever-changing world, so you need to continuously observe your ongoing rhythms and any unexpected events or emerging habits that may affect your gut distress. Second, becoming aware of the interconnections of your life is itself a form of healing, plus it prepares you for the next stage of self-empowerment in your CORE program, restoring your balance.

PART III

Restore

At this point, you have centered and observed. You have gained new skills, perspective, and self-awareness. You are now ready to restore.

Restore *is an action verb not used enough in health care. The most common verbs seem to be* fix *or* fight. *We hear them often, as in,* "Please fix me" *or* "Help me fight this problem."

Fix *is a good verb for healing only if we human beings are machines.* Fix *means* "brokenness." Fix *means* "passivity." Fix *means* "recovery comes from an outside expert who has all the knowledge and does all the work."

Fight *is a good verb for gut distress if there is in fact a battle, an enemy, a war. But what if gut distress results from imbalance and disharmony?*

Restore *is much more appropriate for people who suffer gut distress.*

Consider your recovery from a skinned knee or a cut finger. Healing is spontaneous. Our only challenge is to support our natural capacities to heal. This is what we mean by restore: *to provide or create the right environment, the right ecology, so that healing naturally follows.*

In part III, Restore, you will learn of the latest findings in science relevant to recovery from gut distress. You will learn of the ecologies you can rebalance. You will learn of the brain-gut pathways you can rewire. Most importantly, you will learn of the powers you can unleash. The secret ingredient is you.

7

Ecological Rebalancing for Inner Peace

Happiness is not a matter of intensity but of balance, order, rhythm and harmony.

—Thomas Merton, Trappist monk

B y now, you have come to realize through your observations that your gut is not merely the center of your body/mind system, but it is also the center of your job, your home, your family, your diet, all of your relationships, and your entire life. You are connected to every facet of your life, and each part of it can have an effect on your gut, either directly or indirectly.

Think of all these different elements as little planets orbiting your body/mind system, and your gut is the center of gravity. If something goes wrong with one world—say, your realm of sleep— then it tends to pull the other worlds out of their proper orbits. This stress creates too much pull on your gut, which responds with distress symptoms to let you know that something is wrong. When any of your planets take too much attention or otherwise cause a drain on energy, it affects your entire solar system—and you'll feel it in your gut.

When planets collide—when your world of work conflicts with the world of your family, for example—the collision will also have repercussions on your gut. To make sure your worlds orbit around your center harmoniously, keep pulling them into balance. This rebalancing process continues for your entire life—just as a juggler cannot stop his motion, lest all the balls crash to the ground. Your balancing act requires a 360-degree approach, because everything is connected.

Your relationship to everything that affects your body/mind is called your personal ecology. Examples of ecosystems outside the human body include a forest, a desert, and a coral reef. If any element in a forest system changes drastically, such as a drought or the introduction of an invasive bug that kills trees, the ecosystem goes out of balance. In all of nature, ecosystems are in a constant state of flux—they are constantly adapting to changes in the environment.

We begin this journey into your personal ecology in this chapter by introducing you to your inner ecological environment. Then we'll continue it in chapter 8, "Harmonizing Your External Environment," where we'll show you factors that affect your external ecology. Together these chapters will empower you to restore your digestion to a state of comfort and proper function.

The Inner Ecology of the Gut

June 8, 2012, marked the start of a completely new era in medicine. On that day, the most respected science journal in the world, *Science*, a journal in which Nobel Prize–winning scientists often seek to have their articles published, devoted an entire issue to the topic of the human microbiome, the bacterial ecosystem in and on our body.

The human inner ecosystem, the 10 to 100 trillion bacterial cells in our gut, is now officially recognized as a crucial factor in our overall health. As a reader of this book with a vested interest in your gut, you have a front row seat to witness this revolutionary change in understanding health and healing.

We have known for decades that our inner bacteria ferment otherwise indigestible foods for us; produce vitamins K, B_{12}, niacin, biotin, and folate for us; metabolize hormones and medications for us; and even feed us. What is completely new and revolutionary is that now we know our bacteria actively affect our energy, our metabolism, our mood, our behavior, and our immune function. Yes, these trillions of gut bacteria play significant roles in regulating the 1 trillion human cells that we call ourselves. Thank goodness our body cells are much larger. We may be outnumbered, but we are not outmatched. However, we must tread carefully.

Under normal conditions, we humans work closely and harmoniously with our bacterial friends. When our inner ecosystem is balanced, we are healthy. We can fend off pathogens including bacteria, yeast, and parasites. Our bacteria help us *Bacteria help us fend off toxins from our gut.* fend off toxins from our gut and help maintain an intact barrier between our body and our gut's contents.

Disruptions in our inner ecology, however, have a great deal to do with chronic GI distress. Consider these cases that illustrate the impact of ecological disruption in gut health and Ecological Rebalancing for well-being.

Keith's Post-Infectious Irritable Bowel

Keith used to be a very healthy outdoor enthusiast. That changed, unfortunately, nearly twenty years ago during a two-week hike in Tanzania. Before his trek, he dined in a small city and happened to drink the local water. He then became very ill, suffering from severe abdominal pain, diarrhea, nausea, and dizziness. In a small remote village, someone gave him an unknown medicine for three days.

After he returned home, the symptoms persisted without a clear diagnosis despite his being tested for parasites, among other things. He was treated presumptively for giardia infection, with no effect. Keith developed intermittent constipation, and his cramps became so painful that it was hard to sleep. Later he developed serious sinus infections and was treated with oral antibiotics.

Eventually, he saw Dr. Plotnikoff, who analyzed his medical history. He noted that Keith had a convergence of risk factors that affected the health of his gut's inner ecology. They included several months of pre-trip treatment for acne with a highly toxic medication linked to inflammatory bowel disease. His risk factors also included previously undocumented immunological reactivity to gluten and wheat products. To top it off, antibiotics for infectious diarrhea are known risk factors for post-infectious irritable bowel syndrome.

You can probably imagine what all these medicines did to the trillions of beneficial bacteria in Keith's gut. Testing revealed excessive levels of yeast in his stool, so in addition to wheat avoidance, Dr. Plotnikoff prescribed a gluten-free diet, an antifungal diet, probiotics, and an antifungal medication (Nystatin). Keith reported that he felt the best he had in years while taking the Nystatin, but symptoms returned when he got lax with his diet. Keith later made gains after an elimination diet.

Nicole's Functional to Dysfunctional Bowel Disorder

Nicole's five years of gut-related misery began when she developed heartburn at an arts festival. She was prescribed antacids, but they didn't help. After a year and a half of these medications, she began experiencing significant vomiting, diarrhea, and upset stomach throughout the day. Nicole got sick every time she ate. GI specialists gave her complete evaluations, but they could not explain her symptoms. Every test she took came out negative.

She tried medications for inflammatory bowel disease, stomach acid reduction, constipation, and nausea—but nothing worked. The gastrointestinal specialists finally gave her a diagnosis of functional bowel disorder. They told her that they believed her GI symptoms were due to depression and stress.

Actually, this news made her depressed. She gave up on Western medicine and retreated from life. She got worried after losing twenty pounds, despite trying to eat as much as possible. This absolutely required a medical evaluation. When she first saw Dr. Plotnikoff, Nicole described her symptoms as varying from tolerable to unbearable. Formal evaluation of her intestinal ecology demonstrated complete absence of two crucial types of beneficial bacteria, lactobacilli and bifidobacterium, the same bacteria found in probiotic supplements. Additionally, her gut ecology included two potential pathogens, and one true pathogen.

Dr. Plotnikoff prescribed antibiotics to kill the bad bacteria, prebiotics to nourish good bacteria, and probiotics to introduce new beneficial bacteria to her gut's ecosystem. A more harmonious ecological balance was restored fairly quickly, and Nicole's symptoms were markedly reduced. Within three months Nicole had nearly returned to her baseline weight!

These cases, although extreme and requiring special tests, demonstrate the power of our intestinal ecology to wreak havoc with our gut—or to help it out. Severe intestinal infections spark IBS about six times more frequently than other risk factors do. Those odds rise when the infection coincides with episodes of physical stress, such as prolonged fever, anxiety, or depression—and even mountain climbing. Researchers have found that chronic gut sufferers tend to have significantly unbalanced microbiomes. These distorted microbiomes appear to foster low-grade inflammation in our intestinal wall and other mucous-lined areas, increase our sensations of pain, and foul up the way the intestinal nervous system coordinates digestion.

In this chapter, we will take this insight further so you can understand what triggers—and what can heal—GI distress. This overview of the inner ecosystem will provide you with the newest medical insights.

Introduction to the Microbiome— "We Are Not Alone"

You have already become acquainted with two of the main players in our internal ecology: the digestive tract (stomach and intestines) and the intestinal (enteric) nervous system—the second brain. The third major player, called the microbiome, almost qualifies as a third brain, given its autonomy and ability to affect our well-being. Together, this inner ecosystem has power over our metabolism, mood, energy, immune system, and more. Although this third entity operates as a system, it is not a single thing but a colony within us of some 10 to 100 trillion bacteria along with lesser numbers of viruses and fungi. That's between ten and one hundred times as many microbes as there are human cells in our body. As we previously mentioned, the microbes are much smaller than our human cells and weigh in total less than five pounds.

The study of our microbiome is at the cutting edge of scientific discovery. The greatest concentration of bacteria is in our gut—primarily in our large intestine—where several pounds of bacteria dwell. Because most of us were raised thinking that bacteria are the enemy, two questions may come to mind: What the heck are these organisms doing in us? And why don't we just kill them?

The answer is that not all bacteria are bad. On the contrary, our life depends upon these bacteria. We need these organisms for our own digestion, immune function, mood, and overall wellness—including gut wellness. If we kill even some of them, it throws off our inner ecology. This happens all the time when we take antibiotics and other interventions. Sometimes our body can recover easily, but sometimes it can't. Disrupt the ecology enough and—whoops—suffering may follow.

Instead of fighting bacteria, we need to work with them and support those that will help us as best we can. If we care for them, they will care for us. When they are happy, we are happy. Not surprisingly, the opposite is also true. When we get stressed out, the harmonious ecology is disrupted, which can result in—you guessed it—intestinal distress. Additionally, many opportunistic bacteria have special stress sensors that turn on their genes for disease-causing powers only when people are most stressed and most vulnerable. In other words, any of the five forms of stress can increase our vulnerability to illness. More than anyone else, gut sufferers need to be on the friendly side of the trillions of germs in our gut.

The Role of the Microbiome in Our Internal Ecosystem

We have a symbiotic relationship with our microbiome: we help each other. We give the shelter of our body and a steady supply of food that we send traveling down our digestive system. In return, our bacterial guests do a lot of important work for us, such as breaking down otherwise indigestible food; metabolizing hormones and steroids; producing vitamins, antioxidants, anti-inflammatory compounds, and pain-relieving factors; and strengthening our immune system. They literally feed the cells that line our gut. Our friendly bacteria also protect us against invasive or dangerous microbes. And if that's not enough, we now know that our intestinal bacteria have some capacity to regulate our mood, behavior, and metabolism.

As many as one thousand species of bacteria and other organisms dwell inside us, and it seems that each of us has quite a different mix of them. Our microbiomes are highly individualized, like fingerprints, and they are constantly changing. Even identical twins do not have identical intestinal ecologies.

Our intimate relationship with these microbes is an evolutionary inheritance. All animals have them, and we quickly acquire them at birth. But we are still learning the many ways they benefit us. For example, scientists have recently discovered that a certain bacteria that aids in digesting milk and is normally found in the gut suddenly starts flourishing in a pregnant woman's vagina as she approaches delivery. As the baby is born, it picks up that bacteria from the vaginal walls, and thus has an ally to help digest breast milk. Who knew? But now we also know that babies delivered through C-sections don't pick up those bacteria but collect others instead. Similar stories exist for breast milk versus bottle feeding. Breast milk provides passive protection with natural antibacterial components such as immunoglobulin A (IgA) and promotes colonization of the gut with the beneficial bifidobacterial species. These facts are significant. What we acquire soon after birth, according to studies of laboratory animals, affects the development and function of our gastrointestinal, immune, neuroendocrine, and metabolic systems. For this reason, birth by C-section, time spent in the hospital, length of time breast fed vs. bottle fed, and the amount of antibiotics taken during childhood are potentially very important aspects of your medical history, especially if you have gut distress.

Our intimate relationship with these microbes is an evolutionary inheritance.

Just how powerful is intestinal ecology in our everyday life? Consider how the microbiome may regulate metabolism and mood.

Beneficial Bacteria and Weight Loss

Since 2004, research scientists from around the world have demonstrated that simply changing the intestinal bacteria in mice significantly changes their tendency to gain or lose weight. Fat mice become thin, and thin mice become fat without a change in diet. The only difference is their intestinal ecologies. A world-famous professor of microbiology, Zhao Liping, tried this on himself. He singled out a bacteria (*Faecalibacterium prausnitzii*) that was prevalent in the thin mice, and then he switched to a diet conducive to the growth of that bacteria. (Foods which encourage and support the growth of friendly bacteria are termed *prebiotics* and are discussed later in this chapter.)

For two years, Dr. Zhao ate lots of prebiotics including fermented vegetables and whole grains. The result: *Without any other effort*, he lost forty-five pounds! Not only that, but he also lowered his blood pressure, heart rate, and cholesterol levels. Throughout this time, his laboratory tested his intestinal ecology and documented that the prevalence of the "anti-obesity" bacteria had gone from undetectable to nearly 15 percent of all the bacteria in his gut.

Beneficial Bacteria, Stress, and Mood

Many animal studies and one human study say that an imbalanced population of bacteria in our gut can actually make us anxious. We now know that the acquired microbiome around the time of birth in animals regulates the stress response threshold in our neurohormonal system, specifically the HPA axis discussed in earlier chapters. This means that our gut ecology as small children can affect the threshold at which our system starts to produce stress hormones. If the beneficial bacteria dominate our intestines when we are quite young, our stress response develops normally. But if the bad bacteria hold sway and cause a lower stress threshold, then we could become too anxious. If the threshold is pushed too high, we won't react quickly enough in dangerous situations.

In the largest human study so far on the relation between intestinal bacteria and stress, researchers gave normal adult volunteers either a probiotic (with a lactobacillus species and a bifidobacterium species) or a placebo for thirty days. The test results showed a significant reduction in psychological distress among those who took the probiotics. Results indicated lower rates of stress hormone production, depression, anger, and bodily symptoms, and an increase in problem-solving skills.

The role of our microbiome in regulating both the stress response and mood is one of several links between Ecological Rebalancing and Neurohormonal Retraining found in the CORE program.

The potential impact of our intestinal ecology on our health and wellness is mind boggling. With so much hanging in the balance, we need to understand what sorts of things can turn our microbiome from friend to enemy.

Our Fickle "Friends"

When it is healthy and happy, our microbiome is our friend. The combined power of a balanced brain-gut-microbiome axis can help control movements of the bowel, various hormonal secretions, blood flow, intestinal permeability, immune activity, intestinal pain levels, the processing of nutrients, and even the quality of our mood! But this is a highly sensitive relationship. Any imbalance can negatively affect any or all of these processes.

There are several reasons things can go bad. First of all, while most of the bacteria are friendly, there are plenty of dangerous germs and viruses ready to overtake our intestines. These dangerous microbes are kept under control—literally outnumbered—by the predominantly good bacteria. We don't get sick because the number of bad bacteria isn't large enough to infect us. It's like a game of musical chairs in which there is such an overwhelming number of good bacteria, the bad ones never get to sit down.

However, if anything happens to reduce the number of good bacteria, an opportunity opens for pathogens to flourish or even take over. The bad germs in our gut could take advantage of the situation, or it could be new germs that enter our system because we forgot to wash our hands. As suggested in Keith's story, the results can be quite problematic. The most serious example of this today is the epidemic of potentially fatal *Clostridium difficile* (C. diff) infections in health care institutions. Use of antibiotics and antacids are potential causes of this ecological imbalance.

A second danger is that although our gut lining, with its tightly connected cells, is a strong protective barrier, it is only one cell thick. Total internal surface area of the intestines is about the size of a tennis court. That is a lot of space to cover and a lot of threats to hold off and manage with a lining just one cell thick. That is why gut wall integrity is such a significant issue. With enough of any of the five forms of stress, the intestinal wall can become leaky, resulting in the immune system going into over-drive to manage bacteria that break though this lining.

The Frontlines of Defense

The intestines require a large surface area to absorb the maximum amount of nutrients during digestion. This makes a pretty big target for the trillions of harmful bacteria and pathogens in the gut that would love to wreak havoc on the other side. Our body has three immediate defenses to prevent the bad stuff from leaking through:

1. Physical: Single layer of cells covered with a layer of mucous

2. Chemical: Natural antibiotics made of amino acids (antimicrobial peptides)

3. Immune: Antibodies (IgA)

All three defenses can be threatened by the overgrowth of potential pathogens, infections, inflammation, and contact with reactive foods. The good news is that we can do things to strengthen each layer and promote intestinal healing.

The physical layer can be supported with supplements. Try taking the amino acid glutamine (up to 3,500 milligrams a day), N-acetyl-cysteine (NAC, 500 milligrams twice a day), zinc (20 milligrams a day), or quercetin (500 milligrams twice a day). Additional herbal therapies exist, too. Talk to an informed health care professional to determine the best plan to support the physical layer in your gut.

The chemical layer can be supported with supplemental vitamin D. Have your blood level tested, and talk with your physician about the appropriate dosage of supplementation. The ideal blood level of 25-OH-vitamin D is 40–60 ng/ml.

The immune layer can be supported with body/mind practices that lower stress hormones, plus pre- and probiotic supplements.

Dietary Sources of Intestine-Healing Agents

Add these foods to your diet to supplement your body's nutrients naturally:

- L-glutamine: beef, chicken, fish, eggs, dairy products, wheat, cabbage, beets, beans, spinach, and parsley
- Cysteine (not NAC): animal proteins, eggs, cottage cheese, yogurt, broccoli, red peppers, and onions as well as bananas, garlic, soy beans, linseed, and wheat germ
- Zinc: shellfish, red meats, and wheat germ/bran as well as sesame, poppy, pumpkin, and sunflower seeds
- Quercetin: black and green tea, apples, citrus fruit, onions, red grapes, tomatoes, broccoli, leafy green vegetables, and berries

A third danger is that foods or toxins can leak past the intestinal lining. For this reason, leaky guts appear to be a significant factor in both adverse food reactivity (IgE or IgG) and autoimmune illnesses including inflammatory bowel disease, type I diabetes, and several forms of arthritis. There are also strong links to many skin conditions including psoriasis as well as certain types of cancer. Adverse food reactivity appears to be related to food coming in contact with the immune system that guards the gut. This doesn't happen under normal circumstances, but people with significant gut distress, such as Keith, can have significant immunologic responses to many foods.

Heavy overgrowth of one bacterial strain is problematic if it comes into contact with the immune system because it can cause autoimmune illnesses in susceptible people. These devious bacteria imitate biological markers on normal human cells through a process called molecular mimicry. When the immune system forms antibodies to attack the harmful leaked gut bacteria, it gets confused by the biological markers, so it accidentally kills human cells. These post-infectious autoimmune disorders include inflammatory bowel disease and a form of obsessive-compulsive disorder in children.

Bad News/Good News

Both internal and external forces can upset the healthy balance of our intestinal ecosystem. When that happens, our symbiotic relationship of mutual back scratching with our microbiome devolves into dysbiosis—a dysfunctional ecological imbalance. Our inner peace becomes a war zone:

1. Any of the five forms of stress can disrupt our intestinal ecology.

2. Any of the five forms of stress can undermine our gut wall's protection.

3. Altered gut ecology + altered gut wall integrity = bad news

The good news is that you can revive your population of friendly bacteria in several ways:

1. Reducing/preventing stress with body/mind practices

2. Strengthening gut wall integrity with diet, supplements, and body/mind practices

3. Minimizing adverse food reactivity

4. Adopting a prebiotic-rich diet

5. Supplementing with probiotics

We have already taken a look at how stress management and strengthening the gut wall help to keep the beneficial bacteria dominant. Let's look now at three more ways to restore the harmony of our microbiome.

Minimizing Adverse Food Reactivity: Elimination and Rotation Diets

One of the most effective means of determining which foods can trigger symptoms is to eliminate all suspect foods for a minimum of two weeks (elimination diet) and then to carry out challenges one at a time (rotation diet). The focus of the elimination diet should be to identify any foods that appear to cause you intestinal distress—either allergic or adverse reactions. Elimination and rotation diets are simple trial-and-error strategies to use if your symptoms can't otherwise be explained.

The five most likely suspects are:

1. Wheat

2. Dairy

3. Corn

4. Nightshades (tomatoes, potatoes, eggplants, peppers)

5. Legumes (peas, beans, and peanuts)

Although uncommon, some people may be very reactive to citrus foods.

With such a comprehensive list, you might ask, "What's left to eat?" This is where your food and symptom diary is helpful. You may be able to see right away that some of these foods don't bother you and so do not need to be eliminated. You may be able to pinpoint the most likely suspects for negative reactions with your body. Consider also the role of potentially cross-reactive foods (see pages 67–68). Regretfully, you might find that some of your favorite foods that you eat most often may be the foods that bother you the most. If the elimination diet seems overwhelming, contact an informed nutritionist or dietician who can develop a customized plan for you.

Check *www.trustyourgutbook.com* for additional resources. These include perspectives on adverse food reactivity from traditional Chinese medicine, Ayurvedic medicine, and the contemporary FODMAP dietary restrictions.

After clearing potentially problematic foods via the elimination diet, the rotation diet slowly puts foods back in. We recommend introducing one element of one food group per week (wheat, dairy, corn, nightshades, legumes, or citrus). The first two days of the week, we recommend aiming for three daily servings of the food group and then resuming a more normal intake pattern. During this time, center and observe your body's reactions. Continuing your food and symptom diary can be helpful here as well. Your body is far more sensitive than the latest technology for testing adverse food reactivity. You'll surely notice if your body says, "Yes!" or "No!" to any of the rotated foods.

If your body says, "Yes, great!" You are ready for another challenge in five days. If your body says, "No, stop!" then stop the food and wait twelve days before introducing a new food.

Stephen Helps Himself

Stephen was an art administrator who had suffered gut distress for years and had no idea of its cause. Then he happened to talk with his sister who told him she used to have the same problem, but it went away when she eliminated gluten from her diet.

Stephen followed suit, and sure enough, his symptoms vanished. Now that he shuns gluten, he is amazed at how good he feels. He said he had forgotten what it was like to feel that way, because his suffering had become normal for him.

Adopting a Prebiotic-Rich Diet: Supporting the Growth of Healthy Bacteria

Prebiotics are edible but indigestible products that feed our friendly bacteria. In other words, you can use them to nourish a healthy microbiome and thereby contribute to your balanced inner ecosystem. Prebiotics can't be simply identified with foods; more accurately, various foods *contain* prebiotics. But because they are in foods, prebiotics can easily be worked into your daily diet.

Prebiotics come in many forms:

1. Fermented vegetables and cultured foods:
   - Curtido from multiple vegetables; sauerkraut and kimchi from cabbage; cottage cheese, yogurt, kefir, lassi, and leban from dairy; tofu, miso, natto, shoyu, tamari, and tempeh from soy

2. Foods containing prebiotics:
   - Vegetables and fruits: artichokes, asparagus, bananas, burdock root, chicory, dandelion greens, eggplant, garlic, honey, Jerusalem artichokes, leeks, onions, and legumes including peas and soybeans

3. All soluble and insoluble fiber sources:
   - carrots, apples, oranges, barley, nuts, flaxseed, beans and peas; the beta-glucans found in baker's yeast and medicinal mushrooms as well as psyllium husk

4. Commercial prebiotic supplements, including but not limited to products with these ingredients:
   - inulin-type fructans, fructooligosaccharides (FOS), galactooligosaccharides (GOS), and larch arabinogalactan

Supplementing with Probiotics: Friendly Bacteria

Probiotics are defined by the World Health Organization as "live microorganisms which when administered in adequate amounts confer a health benefit on the host." Lactobacillus and bifidobacterium species are the most commonly studied and used probiotics today. However, *Saccharomyces boulardii,* a beneficial yeast, can be used by people without known yeast allergies or reactivity.

Non-pasteurized cultured or fermented foods—yogurt, kefir, or sauerkraut, for example—are reasonable dietary sources of probiotics. Commercial probiotic supplements are readily available at health food stores and specialized grocery stores including co-ops. Ideally, the product should have at least 20 billion CFUs for the entire shelf life and multiple lactobacilli and bifidobacterial species. These should be taken with cool, unchlorinated water (chlorine in tap water kills bacteria), at least thirty minutes away from warm food or drink. (In other words, taking probiotics with a cup of tea or bowl of oatmeal only makes for an expensive breakfast.)

Since the Human Microbiome Project report in 2012, it is reasonable to predict that much more research will be conducted in the field of probiotics. The safety of probiotic supplementation has been well established for most people. Exceptions include people with critically severe illnesses and significantly compromised immune systems. And, clinically, they obviously play a crucial role in recovery from gastrointestinal distress.

So why are pre- and probiotics not routinely recommended by physicians? There are many potential answers. Clearly, for population-wide recommendations, we need double-blind, placebo-controlled studies. These require many years of preparation to answer some basic questions: What is the right dose? What is the right combination of bacteria? How frequently and for how long should they be taken? Who is the right patient population to treat? Despite more than ten thousand published studies in peer-reviewed medical literature on probiotics in humans, these foundations for rigorous clinical trials have not yet been met. This means that all patients suffering now have to wait at least another ten to twenty years before the scientific community says that probiotic supplementation is scientifically supported.

However, you are free to see if the safe and potentially helpful products available today work for you. Some people can be sensitive or reactive to some probiotics. If this is the case for you, we recommend starting with a low dose of a single bacterial species (1 billion CFUs or less). Then you can slowly increase the dose and expand the varieties of friendly bacteria. The goal would be to achieve a daily dose of at least 20 billion CFUs of multiple species.

FODMAP Diet

Researchers at Monash University in Melbourne, Australia, developed this diet as a means to minimize painful bloating, gas, and other symptoms in IBS sufferers by reducing short-chain and fermentable carbohydrates. FODMAP is an acronym for the names of the offending carbohydrates: Fermentable, Oligosaccharides, Disaccharides, Monosaccharides, and Polyols. Accordingly, the recommendation is for a *low* FODMAP diet. This is a relatively new diet, but it has shown to be successful in a number of studies. The theory is that these carbohydrates are difficult to digest, so they ferment in the intestines causing an excess of liquid and gas, not to mention a random boost to the gut flora population.

High FODMAP foods to be avoided are:

- High-fructose fruits such as apples, cherries, pears, peaches, and watermelon

- Lactose-containing foods like milk, soft cheeses, and yogurt

- All sorts of legumes

- High-fructose corn syrup

- Artificial sugars like sorbitol

- Grains such as wheat and rye

- A wide range of vegetables (avocado, beets, broccoli, brussels sprouts, cabbage, onions, and peas)

The low FODMAP foods that are encouraged include:

- Low-fructose fruits like bananas, grapefruit, lemons, limes, and oranges

- Maple sugar and white sugar (glucose)

- Vegetables ranging from peppers, carrots, and celery to eggplant and tomato

- Oatmeal and spelt grains

For more information, go to *www.trustyourgutbook.com.*

Traditional East Asian Diet

China and other Asian countries have had an approach to healing for thousands of years that lies outside the paradigm of Western science and medicine. One element relevant to gut distress is that different foods have intrinsic properties: they can be hot or cold or damp. We are not talking about temperature, but rather the nature of the food itself.

According to traditional thinking, everything in the world is composed of two complementary forms of energy: yin and yang. Just as cosmic harmony is the proper balance of yin and yang, proper health in an individual depends on a balance of hot (yang) and cold (yin). Any excess of hot or cold or damp creates an unhealthy imbalance that needs to be adjusted.

If the body has too much hot energy, as assessed by a trained practitioner, it needs to be cooled and/or heat needs to be released. Melons, cucumbers and tomatoes can be considered cooling foods.

Likewise, if the body is too cold, then it may need to be warmed. Common warming herbs include ginger and cinnamon.

Dairy and greasy foods are considered very damp. In people with impaired digestion, chronic fatigue, or fibromyalgia, eating damp food can cause phlegm in the throat and/or mucus in the stool.

Any form of gut distress can follow from dietary imbalances of hot, cold, or damp foods. The food and symptom diary can provide insight into foods that trigger gut reactivity. For further information on this topic, see *www.trustyourgutbook.com*.

Ayurvedic Diet

Ayurveda, the ancient medicine of India, defines health as the proper balance of body, mind, and spirit. Although its belief structure is also outside of the contemporary paradigm of Western medicine, it has the same metaphysics as pre-modern science or alchemy in the West, which held that the world is composed of five elements: water, earth, air, fire, and sky (ether).

Everything and everyone is composed of combinations of these five elements. People are said to have three different sorts of metabolic personalities, or *doshas*. A *vata dosha* is a combination of space and air; a *pitta dosha* is a combination of fire and water; the *kapha dosha* is a combination of water and earth. Different types of people are pacified with certain foods, but those same foods would not help balance other types of doshas. You have to find your dosha identity and then seek the foods that pacify that type. Beware of mismatched foods, because they can adversely affect your mood, behavior, and experience of gastrointestinal distress. Learn more at *www.trustyourgutbook.com*.

8

Harmonizing Your External Environment

The best and safest thing is to keep a balance in your life,
acknowledge the great powers around us and in us.

—Euripides, Greek playwright

O ur outer ecology is our network of relationships to all people and things outside of us. How we manage our outer environmental balance directly affects the crucial inner ecology of the gut. The goal is for you to establish a balanced inner ecology but also a balanced outer ecology: observing and skillfully balancing the demands of work, productivity, and helping others on one hand, and the needs for rest, recreation, quiet time, and rejuvenation on the other.

How Important Is This Issue?

Every day, we see well-intentioned gut sufferers inadvertently sabotage their digestive system by failing to establish a balance between work and rest, time for others and a time for self.

When work and family demands conflict with balance and rest, the result can be increased gut pain, cramping, bloating, diarrhea, constipation, and other digestive distress. These conflicting demands also can increase physical stress, dysregulate the nervous system, and boost anxiety and depression while decreasing work performance, job satisfaction, and your general well-being.

Selfless Megan

Megan was an intensive care nurse at a major university medical center. She had struggled for more than four years with chronic cramping, bloating, indigestion, and diarrhea. As she started her CORE program, she became aware that she could reduce her distress through rest, periods of deep breathing, daily physical exercise, and stretching.

When she came to her next appointment, Dr. Weisberg asked how she was progressing with her self-care. She said she had done very little of it and that her symptoms had increased. When asked why she didn't take better care of herself, Megan replied: "I'd really like to, but there's no time! I face all these demands at work, and I can't get a minute to myself. I usually work through lunch. When I get home, my two daughters need me because they haven't seen me all day. What kind of mom would I be if I ignored them?"

We're not the only ones to notice the importance of this. The Institute of Medicine, in a recent annual meeting, observed that the United States does very well in the treatment of acute illnesses (such as heart attack, stroke, or acute cancer treatment). Where our system fares poorly is in the treatment of chronic illnesses—including conditions like IBS and chronic digestive distress. The conclusion was that our medical system is weak in motivating patients to be more actively involved in their self-care. Learning to establish balance among the various demands in your life is a vital step toward improved self-care and in healing your digestive distress.

What Is Lifestyle Balance?

Every day, you perform in multiple roles as employee, employer, friend, supervisor, parent, spouse, partner, board member, congregant, team member, coach, sibling, son or daughter, and so on. If you're like most people, you want to be responsible in every role you perform, but it is impossible to find enough time *Learning to establish balance among the various demands in your life is a vital step toward healing your digestive distress.* for everything. Life has gotten too busy. This leads to a constant tug-of-war among the external demands for your time and attention. Lifestyle balance is the effort to ensure that your personal needs aren't swallowed up by your obligations, and vice versa.

The dilemma for gut sufferers is that they don't always take care of themselves, even though they know they should. Everyone knows they need to carve out time in their schedule for rest, balance, quiet time, recreation, and unstructured time to be quiet and grounded. How do you achieve a sustainable lifestyle balance? Focus on establishing a healthy mix of physical, mental, emotional, and social stimulation, with no one area dominating the others. Balancing the demands of others and self is key to achieving harmony between your external ecology and your internal ecology.

Factors That Affect Your Lifestyle Balance

To evaluate your own self-care and lifestyle balance, look no further than your observations from chapters 3, 4, and 5 (about stress, diet, and sleep). Do you arrange at least one day off from work each week? At the end of your workday, are you able to disengage and leave work pressures behind, or do you feel mentally burdened with work problems during your evenings and weekends? Do you eat a balanced diet and drink enough water? Do you allow time for exercise, recreation, and support from others? Do you set limits when the demands of others for your time become excessive? How balanced are you?

Once you review the problems and solutions concerning stress, diet, and sleep, you are prepared to improve the balance of your internal and external ecology.

Ways to Improve Balance Between Inner and Outer Demands

Follow these tips to restore balance in your life.

Observe the Problem Through Self-Awareness

As you'll recall, you can't change a problem if you don't see it. The capacity to observe and be aware of imbalanced areas in your life is the prerequisite for making improvements in your inner and outer ecologies. Many gut sufferers are oblivious to life imbalances until a flare-up of IBS symptoms demands their attention. The goal is to recognize areas and incidents of imbalance and to take corrective action *before* your troubled digestion becomes the messenger!

Before you can correct a lifestyle imbalance, you must sense the imbalance. Various sensations and emotional responses are clues. How can you tell? First of all, a flare-up of gut distress can be a signal for you to ponder what might be out of balance. Other clues are chronic or acute feelings of fatigue, impatience, anger, sadness, distractibility, anxiety, or tension. An imbalanced lifestyle may also be indicated by sleep disturbance, catching colds more frequently, or withdrawing from social contacts.

Sometimes, simply through the practice of focusing on your gut sensation without trying to change it (Neurohormonal Retraining), you may get a gut feeling about a lifestyle imbalance that needs attention.

Incorporate Self-Care into Your Life

Once you start paying closer attention to your sensory and emotional early warning systems, you can begin bolstering your self-care practices. Self-care needs can be numerous and varied. Try some of these:

- Sleep well: Allow enough time for sleep and address any obstacles to sleeping well, as addressed in chapter 5.

- Eat well: Ensure proper nutrition, as addressed in chapter 7.

- Exercise: Establish an exercise routine, either on your own or with others. This may include walking, running, bicycling, playing volleyball, bowling, playing tennis, swimming, dancing, weight lifting, or a variety of other activities. Find an activity that you enjoy, and make sure you start slow so you can reap the benefits of regular participation. Healthy exercise is self-reinforcing: it feels good, encouraging you to do it more often.

- Allocate time for outside interests: You may benefit from finding a new hobby like painting, sculpting, writing, or playing a musical instrument. Finding optimal stimulation through educational and cultural events improves your overall balance.

- Volunteer: If you have the time, you may find it very enriching and rewarding to serve a cause you feel passionately about. It could be a social service agency, an organization that serves the underprivileged, a school, or a hospital or clinic.

- Take time away: A change of scene can be powerfully rejuvenating, whether it's a two-week vacation or an afternoon drive to a new place.

- Develop social support: People vary in how much time they need to spend with others, so determining your individual needs for interaction and support is crucial. Then, take the time you need to be with friends or colleagues. Get more involved in the social options available through your church, synagogue, or mosque. Nourish and cherish your intimate and primary relationships. Meeting with a psychotherapist can be a way to both assess and meet some of your needs for personal support, too.

- Manage stress: This is another crucial aspect of self-care in the CORE program. Using calming, quieting, and self-soothing strategies is part of this. The methods are numerous and varied, and discussed further in chapters 3 and 10.

- Know your limits: Be firm when people or organizations make demands on your time that interfere with your self-care. Establish boundaries with the assurance that you are entitled to advocate for your needs. You have to be flexible sometimes, depending on the situation, so it's okay if you don't complete everything on your to-do list every day.

- Allocate quiet time: Daily rejuvenation provides you with badly needed rest and reflection time. It will also support your ongoing skill development for self-awareness and self-observation, so crucial to success in your CORE recovery program.

- Cultivate a spiritual practice if this fits your values and beliefs: Whether through a church or synagogue, a meditation or yoga class, or personal guidance from a teacher, spiritual practice can provide emotional, mental, and physical nourishment.

These are only a few suggestions for improving your lifestyle balance. As you improve your skills in self-awareness and self-observation, you will surely add your own ideas to this list.

Jonah Takes a Break

Jonah was a project manager with a local internet company who sought treatment for chronic constipation, bloating, and feelings of sluggishness.

"You know, it's interesting," Jonah commented. "My schedule is so packed that I was convinced there was no time for anything but work and sleep. But even making very small steps can make a big difference. In the middle of my ten-hour workdays, I've started taking a fifteen-minute walk with a few coworkers. It helps a lot just to get outside, move my body, and break the routine. I've also started taking mini-breaks, where I stand up once every hour, take two or three deep breaths, and stretch for about thirty seconds. I can already feel some of the pressure lifted off of me. I think my gut is feeling a little better, too."

Overcoming Obstacles to Achieve Success

Many gut sufferers understand that they need to restore balance to their lives, and they know what they should do. So what's stopping them? Their excuses have become quite familiar to us: "I know I should get more exercise, but I just get too lazy on the weekend," or "I know I should allow some quiet time in my day, but there are just too many tasks to get done," or *Many gut sufferers know what they should do. So what's stopping them?* "I learned a breathing technique that helped my gut feel better, but I just forget to do it."

Let's identify some of the obstacles to achieving lifestyle balance, and review some possible solutions.

Stop Defining Yourself Through Your Job

Our culture highly values vocational success and earning potential. We see patients who work in corporations where the cultural norm is working sixty hours a week. Nights and weekends are just part of the normal workweek. The pressure from both peers and supervisors to be a loyal employee pushes them to put in long hours without regard to their own rest, nutrition, and balance.

Even people outside of corporations tend to define themselves by their job. Some of our patients are self-employed, but they still seem to gauge their sense of self-worth on how they're doing financially. The trouble here is that life balance is often lost in the race to succeed professionally and financially.

Don't be chained to your workplace identity where your self-esteem fluctuates according to your monthly profit and loss statement. You are worth more than that. The self-care activities listed in this chapter will help you see the bigger picture. Occasionally, some brief psychotherapy may help you reestablish a sense of worth that goes beyond your hours of productivity. One of the most powerful motivators for changing this pattern is the attention-grabbing presence of increasingly painful gut distress.

Rachael the Corporate Climber

"It's been my dream for my whole adult life to be a successful litigator, the best in my field," said Rachael, a thirty-eight-year-old attorney. "You do not progress in this business unless you routinely put in twelve- to fourteen-hour workdays and work weekends, too. Sometimes the pressure and pain in my lower gut is so strong, I can barely concentrate on taking a client deposition. But I'm going to work hard enough to become a partner at my law firm, no matter how many hours or trips to the bathroom it takes. I just couldn't live with myself unless I reach this goal."

Pain Stops Phil

When the pain finally got so severe that Phil couldn't continue to work twelve- to thirteen-hour days, Phil finally exclaimed, "Okay, I give up! I can't continue to push myself day and night to raise my productivity numbers. I'm clearly in just too much pain. Maybe I need to listen to what my gut is trying to tell me and slow down a little."

Phil started to discover the intelligent information that gut symptoms have for us, if we're ready and willing to pay attention to them.

Prioritize Self-Care

Some people feel as if they have to attend to their own work responsibilities as well as to the needs of others before they have the right to allocate time to their own needs for rest, rejuvenation, and recovery. Why should you be ranked last?

Sara's Needs Come Last

Sara was in the clinic again due to her ongoing problems with cramping, bloating, chronic constipation, indigestion, and sharp pains in her lower left abdomen. She also suffered from pain-related insomnia. A month earlier, Dr. Weisberg had recommended she take a number of steps to establish better balance in her life. His suggestions included working up to twenty minutes of self-hypnosis twice a day, keeping a sleep diary to observe the patterns affecting her sleep, and practicing a two-minute breathing technique three times a day to calm her autonomic nervous system.

At her next visit, Dr. Weisberg asked how she did with the self-care exercises. Sara replied, "It's been a very difficult time to keep up with all of this. I have my college final exams and needed to devote extra time to studying. My mother has had back problems lately, so she needed me to come over to the house more often so I could help prepare meals and clean up. And my friend is going through a difficult breakup with her boyfriend, so I have to be available for her when she needs me."

Sara rarely followed through with the self-care practices but not because she was lazy, unintelligent, or unorganized. Somehow, the full meaning of her predicament had never really struck her deeply, until one session in which they worked on Neurohormonal Retraining.

Her eyes welled up with tears as she focused closely on the cramping sensation in her lower abdomen. Then she said, "The pain isn't quite as bad now, but I just had a realization. I always work hard to finish all of my responsibilities and get good grades. But I'm also supposed to be available to meet the needs of anyone important in my life. If my mother, my father, my friends, or my fellow students need me, I feel like I'm supposed to respond to their needs first. If I pay attention to my needs before theirs, I feel bad, guilty, and selfish. I guess I don't have permission to take care of myself until everybody else is okay first."

We see this pattern a lot in the clinic. Gut sufferers often feel it is selfish to take time for self-care until the needs of absolutely everyone else have been met. We hereby grant you permission to respond to your own needs! This is an absolute necessity for your healing. You are a priority in your life. Who is in a better position to know what you need than you? Sometimes your need for rest and relaxation is more important than other people's demands.

Dr. Weisberg shared an anecdote with Sara on this point: "When you're on an airplane before takeoff, you know how the flight attendants go through the routine of reciting safety instructions? They mention wearing seatbelts, putting up the tray tables, and the like. At one point they always say, 'In case of a loss of cabin pressure, the oxygen mask will come down. Be sure to secure your own mask before you help someone else.' Why do you think they say that?"

Without missing a beat, Sara replied, "Because you can't help someone else if you don't take care of yourself first!" She immediately recognized the irony of what she'd just blurted out and said, "I guess that's something I need to start doing, isn't it?"

Once again, your own pain and distress can be a wise teacher, reminding you that you have to prioritize your own self-care time. If you take time to relieve your own gut distress, you'll be more fully available to help others.

Improve Interpersonal Boundaries

If you put others' needs before your own, you may also have problems with interpersonal boundaries, and that can worsen your gut distress. Personal boundaries are demarcations that define your identity as separate from others: personal space zones, physical body boundaries, communication boundaries, and emotional and social boundaries relating to family, marriage, friendship, workplace, and authority. And as a gut sufferer, you need good boundaries for establishing lifestyle balance. Your boundaries protect your self-identity, your integrity, your feelings and needs, your goals and values, and your overall sense of well-being.

You need good boundaries for establishing lifestyle balance.

Operating with good interpersonal boundaries means that you protect your personal rights and individuality when you interact with others. Excessive people pleasing or passive submission to others is harmful to your self-esteem and self-identity. *You* get to choose when you defer to the needs of others and when you choose yourself. People with good relationship skills can choose when they want to be social and when they want to be separate. That's what Robert Frost meant when he said, "Good fences make good neighbors."

John Gets the Message

As part of his CORE program, John realized he got tired of constantly responding to text and email messages on his smartphone. He'd get thirty to forty messages every day, many from family members asking him to do errands or to plan family events. He finally had a meltdown on a particularly busy day at work when he was interrupted by over fifty-five text messages from family. "I love my family, and I want to be there for them," he explained, "but this constant barrage of messages makes me feel so stressed. My stomach starts to churn from just anticipating when the next text or email message is coming in!" John finally decided to leave his phone off during the workday, and he told his family to wait until after six pm to contact him. John successfully made a boundary, and it helped alleviate his gut distress.

The following are some strategies that psychologist and coach Dana Gionta, PhD, suggests to help you establish or maintain better interpersonal boundaries.[3]

- Identify your limits: Use your feelings to locate your boundaries. When you feel uncomfortable, resentful, or stressed, these are clues about what you can tolerate and what you can accept. Learn to name and specify your physical, emotional, mental, and time-related limits. Then, try not to let anyone cross that line uninvited.

- Tune in to your feelings and sensations: If you notice discomfort, resentment, or a flare of gut pain during an interpersonal situation, an improved boundary may be needed. When these feelings arise, ask yourself: What is causing this feeling? What bothers me about this interaction? Resentment, anxiety, or physical agitation may be a sign that you are pushing yourself beyond your own limit. Maybe you feel guilty or not entitled to say no. Maybe other people are imposing their expectations, opinions, or values on you.

- Give yourself permission: Fear, guilt, and self-doubt undercut your willpower. You might fear the other person's response if you hold firm and enforce your boundaries. Does saying no to a family member make you feel guilty? Do you think you should say yes because you're a good daughter or son, even though you feel drained or taken advantage of? You might even wonder if you deserve to have boundaries in the first place. Boundaries aren't just a sign of a healthy relationship; they're a sign of self-respect. So give yourself

---

3 *http://psychcentral.com/lib/2011/10-way-to-build-and-preserve-better-boundaries*, reprinted by permission of Dana Gionta, PhD, *www.counselingconnecticut.com*.

permission to set boundaries and stick to them. It's your right as a human being!

- Practice self-awareness: Boundaries are all about honing in on your feelings and honoring them. If you notice yourself slipping and not sustaining your boundaries, get back to noticing your emotions and sensations. Ask yourself: What am I doing—or what is the other person doing—that is eliciting this feeling or sensation in my body? What limit do I need to set to change this feeling?

- Make self-care a priority: Good self-care skills and good boundaries go hand in hand. The more frequently you grant yourself permission to attend to your own needs, the stronger your motivation to set boundaries. Self-care and boundaries mean recognizing the importance of your feelings and sensations and honoring them. These feelings and sensations serve as important cues about your well-being and your basic need to feel happy, calm, and balanced.

As you continue to grow in your ability to observe your life and develop a keener sense of self-awareness of emotions and sensations, you will find it easier to create better lifestyle balance. As a result, your internal and external ecological systems will be more in balance, and your gut will feel better.

9

Neurohormonal Retraining to Rewire the Gut-Brain Connection

Between stimulus and response there is a space. In that space is our power to choose our response. In our response lies our growth and our freedom.

—Viktor Frankl, Austrian psychiatrist

An important step in the process of healing your chronic gut distress is to distinguish what started the symptoms from what keeps them going. Your malady may have been prompted by a traumatic event such as a serious infection, a convergence of stressors, or both. But what makes it continue for months and years is a miscommunication between your brain and your gut. Your brain has become hypersensitive to any gurgle or twitch in the gut, which prompts it to overreact by releasing stress hormones throughout the body as if there was a major emergency. This neurohormonal response results in a tightening of muscles in the gut that causes cramps, pain, diarrhea, constipation, and all the things you've come to hate. How can you stop this cycle of madness?

The answer is that you have to rewire the communication channel between the brain and the gut through Neurohormonal

Retraining. This is a deceptively simple yet effective tool in the fight against chronic gut distress. The reason you've never heard about it before is that it is a new synthesis of treatments. Dr. Weisberg developed it from his many years of experience using hypnosis for physical illness, and it has been adapted here specifically for treatment of gut distress

The key to Neurohormonal Retraining is to take on a patient and curious attitude toward the sensations within your gut. Rather than ignoring your pain or trying to fight it, adopt the middle attitude of just noticing, observing, and *feeling* it. By holding your awareness in this neutral but focused state, you can short-circuit the negative chain of events that brings on another attack of gut distress. Regular practice of this simple technique can significantly reduce or eliminate your chronic pain.

You can rewire the communication channel between the brain and the gut.

How to Do Neurohormonal Retraining

1. Sit in a comfortable and balanced position in a place where you won't be disturbed.

2. Focus on the feelings in your gut and identify how large of an area is affected by bloating, cramping, or pain. You can use your hands to outline the exact area of distress. How big is the feeling?

3. Assign a number to the *intensity* of the feeling on a scale of zero to ten, where ten is the worst pain that you've ever experienced in your life and zero represents no pain at all. How intense is the pain?

4. On the same zero-to-ten scale, rate the *intrusiveness* of the sensation. How much does this sensation bother you? Is it

just a slight irritation or does it bother you so much that you are distracted from daily activities?

5. Focus on the sensation of your gut distress very closely, exactly the way it is. Feel it, don't think about it or worry about it. Let yourself be *curious* about the sensation. Curiosity is a very specific and important outlook for this technique, because it implies paying close, focused attention. At the same time, curiosity requires a certain level of calm and grounded attentiveness. You can't be curious and anxious at the same time. But you can be curious and remain somewhat detached and receptive.

6. Imagine that you are going to pull up a chair and sit right next to the sensation just exactly the way it is, without trying to change it in any way. Tell yourself, "I may not like this feeling, but I know that it is not going to harm me because it's been medically checked out. So for the next few minutes, I'm going to make room for it to be there *exactly* the way it is." Let it be.

7. Give it some time. This step is crucially important. Allow the process to occur. This is when you create new neural connections and reflexes. As you sit with the sensations, you may feel a pull to get distracted. You may notice getting impatient with the exercise and feel the urge to force the pain to stop somehow. Or you may feel compelled to try to relax it away, breathe it away, or visualize it as disappearing. (Interestingly, this is when *trying* to relax actually becomes a distraction to healing.) It is a natural reflex to want to get rid of it; please don't succumb to that desire. This is when you learn to trust the inner wisdom and inner healing resources of your body. To put a twist on the old adage: don't just *do* something—*sit* there! Let this be an interesting experiment as you learn a new response to an old problem.

8. As you sit with the sensation, don't try to rush it. Give it time. It doesn't matter whether it takes two minutes or ten minutes, because your body will let you know how much time is needed. Just pay close attention to your gut feelings and be receptive to your body's cues.

9. By focusing closely and observing the sensation *without judging or analyzing it*, you are activating different parts of your brain than you would if you were to think about it.

10. There will be times when you get impatient with the exercise and just get so mad at the pain that you try to make it disappear. With enough practice, it gets easier to just sit with the sensation and not fight it. Some patients say it feels like breaking through the wall—the way a long-distance runner feels out of breath and then suddenly gets a second wind.

11. Make room for whatever happens. Sometimes the sensations stay about the same in size and intensity, but more often they will move around and fluctuate in intensity. Many people report that the distress becomes smaller and less painful over the course of a session. Sometimes the pains move out of the gut and up into the stomach or chest. They could switch from the right lower abdomen to the left lower abdomen, or even end up on your side.

12. An interesting phenomenon is that sometimes people get a gut feeling in a metaphorical sense. They suddenly get a hunch about something important related to their symptoms. They may realize, "I'm really upset at my friend for ignoring me," or "I'm really exhausted and need some rest," or "I'm in the wrong job; I hate it." When a clue like this emerges, simply pay attention to it. It could lead to some redirection or calming of the gut distress.

13. Don't be surprised if it feels a little awkward or confusing when you first try Neurohormonal Retraining. With practice, you'll be pleasantly surprised to notice not only that your gut

symptoms start to improve but also that you will begin the priceless practice of learning to trust your gut, and you will benefit from its wisdom about your life and needs.

14. Repeat this process whenever your gut pain or distress becomes truly bothersome.

Carrie Reluctantly Tries Neurohormonal Retraining

Carrie tried various treatments for four years to get rid of her post-infectious IBS. But the diarrhea, fatigue, and bloating only got worse. When it started interfering with her work as a set designer at a local theater company—as well as her social plans— she hated it. Carrie began to waver between anger and depression.

She came to Dr. Weisberg, who suggested she do Neurohormonal Retraining. That seemed to touch the wrong button. "If you ask me to pay more attention to these aches and pains, I'm going to be really upset," she said, almost crying.

She was not the first patient who got riled up about being asked to pay attention to her painful sensations. After all, they've spent months and years trying to ignore their distress.

When Carrie's distress got even worse, she softened her attitude. She said, "Let's make a deal. I'll do some of my best relaxation techniques right now. If that doesn't reduce my pain, then I'll try your neurohormonal whatever-it-is."

"Fair enough," Dr. Weisberg replied. For twenty minutes, Carrie did focused breathing, yoga stretches, and self-guided visual imagery of watching the pain disappear. Nothing changed. Then, with great reluctance, Carrie agreed to try the Neurohormonal Retraining technique.

With much encouragement, she finally sat and focused on her intense bloating sensations. For the first few minutes, she tried to force the pain to go away. But eventually, she allowed herself to sit and simply feel the bloating sensation. After about four minutes, her eyes popped open in surprise. "The bloating is almost gone!" she exclaimed. "How can that be?"

Dr. Weisberg explained that her system needed her to stop fighting the symptoms and to begin to befriend and listen to them.

Then a tear came to her eye, and she said, "Something else happened while I focused on the bloating sensation. A wave of relief and recognition came over me. I realized that I get all my sense of value from being a great set designer, and I'm afraid that if I can't perform at the high level of excellence that's expected of me, no one will value me anymore." Then she started sobbing.

This was a valuable insight, courtesy of Carrie's gut wisdom—once she was willing to listen to it. By addressing her fears and doing Neurohormonal Retraining, Carrie's diarrhea, fatigue, and bloating improved dramatically.

Stay with the Cutting Edge

As we mentioned before, as you do Neurohormonal Retraining practice, sometimes the pain, bloating, cramping, or other uncomfortable sensations actually start to change location, intensity, or even quality. You may find this to be surprising or confusing when you first start to do the practice. But after a while, you will get more accustomed to it. Many gut sufferers eventually find this to be quite comforting. Think of it: when you have been struggling with pain in your gut that has been anchored there for years, it can feel a little liberating and hopeful just to stop fighting it.

When the pain moves, stay focused on the part that hurts the most.

One essential principle of Neurohormonal Retraining is that when the pain moves, stay focused on the part that hurts the most. *Stay with the cutting edge.* That means that you may start by focusing your attention on a cramp in the center of your abdomen, and then you find that the most noticeable spot is a deep ache in your left side. After a while, you may notice that the biggest sensation is a tingling in your upper right abdomen. By staying with this cutting edge, you will maximize your inner healing resources for retraining the responses of your abdomen, your nervous system, and your brain. Remember, pain is a call for attention from your body, and the best thing to do is listen to it.

Jeff Rides the Cutting Edge

Jeff was a forty-six-year-old attorney with a long history of chronic diarrhea and cramping and a long list of unsuccessful treatments. Finally, a friend of his who had success with the CORE program sent Jeff to us.

We instructed Jeff in the practice of Neurohormonal Retraining, with the emphasis on staying with the cutting edge of the painful sensation as it moved and changed. He initially focused on a deep ache in the center of his abdomen and rated the intensity of the pain as five on a scale of ten. After two minutes, he noticed that the cutting edge of the distress had migrated to the center of his chest! Without analyzing or trying to steer it, he just refocused his attention to his chest, now just allowing that sensation to be there while he observed it. Three minutes later, he was surprised to find the focus of his attention switched to a moderate ache in his right lower abdomen, near his appendix.

By now, Jeff was becoming more adept at just following the cutting edge of sensation as it migrated around his abdomen and torso. As it kept moving and changing, Jeff gradually became more and more curious about where the most noticeable part of the aching and pain was going to relocate. By the end of the experience, the painful ache in his belly had been reduced to a mild annoyance in the center of his abdomen, rated at an intensity of one out of ten. He said, "Wow—that's crazy how the cutting edge of my gut pain kept moving around. The less I fought it, the more it moved around and then decreased."

Neurohormonal Retraining and State-Dependent Memory

Have you ever had an experience in which a certain song, smell, or place suddenly sparked a full-blown memory from long ago? That happened to Dr. Weisberg recently when a friend of his played the song "Gold Mine," a jazzy, upbeat piece by an a cappella vocal group named Take 6. That was the same song he had been listening to years ago when he flew to Switzerland for the first time. When he heard the song again, all of a sudden, it brought back the sights, sounds, emotions, physical sensations, and even the fragrances of his arrival in Switzerland. As that song played, he was transported back to Switzerland.

This multisensory collage of physiology, images, thoughts, sensations, and emotions is called a state-dependent memory. A collection of feelings and sensations are bundled together so that when you experience one component, all the rest of the memory comes flooding in. These state-dependent memories can occur either in a positive sense, like Dr. Weisberg's Switzerland trip, or in a negative sense. For example, a person with IBS notices a certain gurgling in her gut, and it sparks a full-blown memory of cramping, anxiety, impatience, worries, and physical tension. The gut sensation has become part of a state-dependent memory that is bundled with an entire traumatic episode of severe gut distress. The original event may have been caused by a serious infection or severe ecological imbalance, but this experience was so vivid and traumatic that it became part of a state-dependent memory.

As long as this memory stays alive, you are haunted by the original episode. Dr. Weisberg was not in Switzerland anymore, but he felt like he was. You are not infected anymore, but it feels like you are. The good news is that Neurohormonal Retraining can help unbundle the different aspects of IBS in your state-dependent memory and can provide valuable clues on how to correct this situation.

The neurohormonal approach to pain is quite different from the reflexive response of the average gut sufferer. The reflexive response is kicked off by the activation of a state-dependent memory with all its baggage. The brain is fooled into thinking something serious is happening, and it fires off hormones that ultimately launch a painful gut episode.

Many gut sufferers think they are paying attention to their bloating, cramping, and pain all day, every day. In fact, this is not what usually happens. Instead, they are flooded by all the worries, recollections, and apprehensions of their state-dependent memory. They remember all the events they canceled, they worry about where to find bathrooms, and they start thinking about possible remedies to fix the problem. All these worries and thoughts actually shift their attention away from the pain in less than a minute, but people don't generally realize this. And they don't readily admit to it either.

Jack Sees the Light

Jack, a twenty-five-year-old customer service worker, was quite surprised when we first told him that he probably only paid attention to his sensations for a few seconds.

"Are you kidding me?" he asked incredulously. "This hot, throbbing ache in my lower left abdomen never leaves me alone. Sure, I've had all the tests to rule out other diseases, but the pain is unchanged. They just told me to get some fiber. I feel this pain every waking moment."

We encouraged him to keep a pain diary for a week, a log of his cognitive and emotional reactions to painful gut sensations. When he returned, he confessed he was flabbergasted.

"Unbelievable," he declared. "You guys were right! As soon as I feel the throbbing ache in my left side, my mind immediately shifts into overdrive, and my attention goes to all of my worries and irritations. I get really, really mad. I even grit my teeth. I start thinking ahead into the future, predicting how long these

horrible symptoms are going to last, making me feel worse and worse. I probably don't notice the actual feeling in my left side for more than two or three seconds at a time."

The Natural Tendency to Avoid Pain

Jack did what most people do. We have a natural tendency to fight or avoid feelings of pain. As the pain lingers, we seem to instinctively tighten up the muscles around that area as if to protect it. If you've ever injured your back, you remember how much stiffer you walked in an attempt to minimize the pain, tightening up your entire torso. The same thing happens in chronic abdominal pain. We instinctively tighten muscles and connective tissue around that area. Our brain and nervous system respond to this tension as well, secreting various hormones that affect pain and movement of stool in the intestines.

Many gut sufferers are prescribed medications to try to get rid of the pain or make it more tolerable. Sometimes these drugs can be helpful in the short term to break the pain cycle or to help the people feel less overwhelmed by the discomfort. If it's used in this way, it's beneficial.

The problem arises when these medications are taken more than occasionally. Then, the drug starts to become yet another method for avoiding the pain. If the people use the painkiller chronically, they may develop analgesic rebound, in which the medication actually alters pain circuitry and creates additional pain. Sufferers may develop drug tolerance, in which they require increasingly higher doses to attain the same level of relief. Ultimately, chronic use of painkillers by gut sufferers deprives them of the opportunity to face their sensations and create a sense of control and mastery over them.

People have also become quite adept at ignoring pain by distracting themselves. The old-fashioned approach is to keep busy at work or some other activity. The modern way is to get in front of a screen and watch television or a movie, read Facebook, or mindlessly surf the web. People may also learn to actually dissociate from the sensations in their body, cut their mind off from the discomfort until it becomes so painful that they can't ignore it anymore. In the course of a day, all of these strategies can seem to help us keep our mind off our intestinal troubles.

People are adept at ignoring pain by distracting themselves.

Edna the Avoider

Edna, a sixty-year-old divorced physician, developed her own daily routine for coping with gut pain. "I hate the constipation and bloating," she said. "It seems to increase throughout the day, so by evening, it's much worse than when I woke up. So when I get home, I turn on the television and turn on the stereo. I watch TV until eleven thirty or midnight every night, just so I don't have to be aware of the pain. The worst part comes when I have to turn off the television and get in bed. That's when everything I've tried to avoid catches up with me."

When you live with chronic pain, it begins to shape how you live your life. The pain signals that remain active in your nervous system change how you participate in daily activities. You walk differently because of the increased tension in your muscles. You carry your body differently. Mobility becomes limited. Energy is reduced, and appetite is disturbed.

And it's not only physical. Emotions like anger and despair become part of the mix for people in chronic pain. Emotional reactions such as depression, frustration, anxiety, and shame combine with the fear that the symptoms won't get better. To make matters worse, these difficult emotions can actually help prolong and intensify chronic pain. It becomes a vicious cycle.

Acute pain works differently than chronic pain. Acute pain activates the original alarm that something has gone wrong due to an infection or tissue damage. Chronic pain, however, gets maintained or worsened by the central nervous system when it is affected by environmental stresses at work, at home, in relationships, and so on. The body gets weakened as it adjusts to a more narrow range of activity. These and other prolonged emotional reactions emerge as more painful bodily responses. The initial pain is a legitimate cry for immediate medical attention. Chronic gut pain, once it's been medically evaluated, is a message telling you that your body/mind system is out of balance.

Pain and Conditioning

Chronic pain also involves learning and conditioning of the nervous system. Most people have learned of conditioning through the story of Pavlov's dog. When Pavlov, a Russian scientist, presented a tray of food to his dog, it would salivate. Then he started ringing a bell at each feeding. After several repetitions, he found that ringing the bell alone without the food could cause the dog to salivate. The same is true for humans. The body learns a physiological response to a neutral stimulus or signal, and this is called a conditioned response.

Because pain evokes such strong emotional reactions, conditioned responses to pain occur often. If someone had an unpleasant or painful experience in a physician's or dentist's office, then the mere smell of rubbing alcohol or the sight of someone wearing a white coat can set off a conditioned response of anxiety.

Dentists who work on children know this phenomenon well. This is why the child's first visit to the dentist is simply a get-acquainted session, so he can just come to the office, sit in the chair, and play for a little while. That way, the child's first association with the dentist's office will be relaxation and fun, rather than fear and trepidation. Then when the youngster comes in for dental work, the child handles the unpleasant experience with less conditioned anxiety, tension, and distress.

For a person who has occasional digestive distress, the feeling of bloating, soreness, or abdominal pain may produce some mild concern, but there is an expectation that the pain will soon go away. No big deal. And no conditioned response. However, for the chronic gut sufferer, pain conditioning can cause added distress that intensifies and maintains the symptoms. A feeling of dull pain in the abdomen, bloating, and gas, or even the experience of frequent diarrhea can lead to powerful conditioned reactions of anxiety, depression, frustration, or hopelessness. These conditioned emotional responses can create a cascade of neurohormonal changes that maintain or increase pain and other symptoms.

The lesson here is that the brain, body, and nervous system can learn and change—for worse or for better. This is also very good news, as it opens exciting new possibilities for healing your gut.

The brain, body, and nervous system can learn and change—for worse or for better.

Neuroplasticity

Scientists used to think the brain and all its neural patterns never changed once they were created. Now we are learning that the human brain and nervous system do not stop growing after childhood, but rather they continue to develop, grow, and change throughout life. We can deliberately carry out activities that change our brain! This lifelong flexibility is called neuroplasticity.

Neuroplasticity means the brain is able to reorganize itself by forming new neural connections to serve new functions. These changes in the structure of our brain can be brought on by life experiences, biological agents, thoughts, behaviors, beliefs, emotions, habits, and life lessons. Neuroplasticity allows nerve cells in the brain to compensate for injury and disease. Even stroke victims with brain damage can relearn skills by forming new neural pathways to replace the ones that were destroyed. The good news for us is that we can reprogram the brain's neural and hormonal responses to digestive distress.

Neuroplasticity

Neuroplasticity is the inherent flexibility of nerves to form new connections with other nerves in response to changes in the body/mind system. This fairly recent discovery in neurobiology means that the brain and nervous system can be reprogrammed throughout life.

The bad news is that sometimes these new neural networks cause a lot of pain and discomfort. The good news is that through neuroplasticity, it is possible to stimulate the inherent self-healing resources to help relieve chronic gut distress with techniques like neurohormonal reprogramming. Here are some examples of negative and positive neuroplasticity:

Negative Neuroplasticity

- Chronic abdominal pain leads to changes in the brain and nervous system. Pain signals are amplified, muscles tighten, anxiety and depression can increase.

- Repeated pain signals alter the place in the brain that regulates pain sensations (the amygdala and cingulate cortex), so the sensations keep feeling much more painful than they actually are.

Positive Neuroplasticity

- Stroke victims relearn old skills by rewiring their neural pathways around the damaged spot in the brain.

- Regularly practicing meditation can improve your ability for finely focused attention, due to neuronal growth in the prefrontal cortex of the brain.

- Several studies have shown that the use of various relaxation techniques allows people to recover more quickly from stress-induced physiological arousal because it supports change in brain structure. Less stress = less pain. Good diet, proper exercise, and sufficient rest are all helpful.

- Neurohormonal Retraining, which lets IBS sufferers pay attention to the uncomfortable sensations of pain, bloating, and cramping in their abdomen, uses neuroplasticity to create new neural connections that reduce the pain they feel. The central nervous system can learn to respond more calmly to the uncomfortable signals sent from the intestines to the brain. It turns out that something as basic as learning to sit with an uncomfortable sensation in your abdomen can affect positive neuroplasticity.

How Focusing on a Sensation Changes Physiology

Chronic pain changes the way the brain responds to uncomfortable sensations in the gut, and it increases the natural tendency to avoid pain. Even though your tests ruled out more serious digestive diseases, your emotional brain has learned to interpret each gut sensation as a threat, triggering cascades of stress hormones and neurotransmitter messages that intensify your pain and bloating sensations. You might say that your rational brain got reassuring news, but your limbic brain never got the memo.

As you can see, in order for your painful gut to heal, you need to retrain your brain and nervous system. Recent breakthroughs in neuroplasticity teach us that our brain and nervous system are capable of learning and structural change.

Neurohormonal Retraining works because of the new way that you pay attention to your bloating, cramping, and other gut pains. This curious but passive awareness retrains your brain and nervous system. With practice, you will find it is one of the most powerful tools in your CORE healing toolbox.

Putting It All in Practice

Imagine observing the aches and pains in your gut this very minute by using these techniques. Merely sitting with a sensation and observing it without struggle changes things. By sitting with and paying attention to your gut pain without fighting or struggling with it, you begin a process of very powerful change. Your attentive curiosity creates a sense of separation from the pain. It is as if there are two separate entities. Continued practice replaces fear with trust, and you become ready to call a truce, an end to the adversarial relationship with your gut.

This important exercise sends a new signal to structures deep inside your limbic brain. You are telling the amygdala and related brain structures, "This sensation is no longer a threat." As you repeat this exercise, the limbic brain starts to signal to the nervous system that these sensations of pain, pressure, and bloating are not a cause for alarm. When this happens, there is a change in the cascade of neurotransmitters and neurohormones that communicate with the gut. Our patients are almost always amazed to find how much the symptoms move, change, and quiet down with Neurohormonal Retraining.

There is a saying in pain treatment: sometimes the symptom is the solution. For us, that means that rather than trying to fight or avoid the painful gut sensation, there is actually benefit in paying attention to the sensation. In fact, paying attention to and making room for the painful sensation may actually turn out to be exactly what is needed for healing.

Sometimes the symptom is the solution.

Patients who do this practice sometimes have epiphanies. They suddenly become aware of a thought, an image, an intuition about something important that they need to attend to. People say things like, "I'm realizing that I really don't like the kind of work I do," or "I'm really upset about issues in my marriage," or "I really push too hard and need to find ways to rejuvenate." This is the inherent inner wisdom of the body, always there if we can learn to listen to it. This is called listening to your gut.

Neurohormonal Retraining: *Minding, not Managing,* Your Pain

Managing your pain often means begrudgingly refraining from constantly fighting it. "Just grit your teeth and bear it," this school of thought says. Tough it out. Do your duty, as if you have a sick friend or relative that you have to take responsibility for, whether you like it or not. You have to be careful about what you eat, how much you exert yourself—and don't forget your medications! You resent your pain and fatigue, but you are resigned to it. You are managing to get by.

In contrast, *minding* your pain is a change of attitude, awareness, and perspective. Once all the serious medical problems have been ruled out, you can befriend your body and be less afraid of your pain. Symptoms are sources of intelligent information calling to you about what needs rebalancing in your life. Information about what your body and mind need to reduce the depression and anxiety that is contributing to your seemingly endless cycle of pain and bloating.

Minding your pain addresses the physiology of chronic pain and what is needed to retrain the body and mind. It unleashes the body's inherent capacity for self-healing. This approach undercuts the cycle of chronic pain through positive neuroplasticity. Minding your pain and befriending your body can lead to vastly improved relief from cramping, bloating, and bowel motility problems. You can feel better, more energized, calmer, and more hopeful. Neurohormonal Retraining helps you get there.

Charlotte Befriends Her Gut

Charlotte chuckled. "It's funny," she mused. "It used to be that the last thing I would ever want to do was to sit and pay attention to these bellyaches and pains that I despised and wanted to get rid of in any way possible. But I've achieved so much relief and comfort through Neurohormonal Retraining that now I look forward to focusing on any new gut sensations, even if they're unpleasant or painful at first.

"It almost always helps me feel better, and it's brought me a surprising sense of influence and impact over my symptoms. Sometimes it lets me tune in to intuitive gut feelings, like an inner wisdom revealing to me some of my deepest feelings and needs. It's ironic—the less I try to control these painful sensations, the more control I have over my suffering."

10

CORE Calming Techniques

Smile, breathe, and go slowly.

—Thich Nhat Hanh, Buddhist monk

In nearly every chapter so far, we've told you this was coming, our collection of calming techniques to help you get centered and stay centered so you can carry out the CORE program of self-healing. We will present an array of skills to soothe, comfort, and rebalance your troubled gut. Stress and tension—both physical and emotional—can be causes of intestinal distress as well as barriers to healing. Moving past them can open you up to the amazing healing potential of your body.

You've already learned to observe the various types of stress affecting you. This allows you to prevent tension by avoiding stressful situations, changing your diet, and paying attention to your needs for balancing rest and activity. Neurohormonal Retraining prevents stress by increasing comfort and healing while rewiring the gut-brain communication channel in your body. But what about all the residual stress that has built up and the unavoidable random stress that hits you from all directions every day? For that, you need special calming skills that help to restore the natural body/mind balance and promote your body's natural healing powers.

These techniques should be practiced daily to calm and tone your gut, your mind, and your nervous system. They complement the other CORE skills of observing your body/mind system and practicing Neurohormonal Retraining. These days, you can find hundreds of stress reduction techniques in books and online, but the methods presented here are selected to specifically help people with gut distress:

- Hypnosis
- Breathing techniques
- Meditation
- Self-massage

Hypnosis

Hypnosis is one of the most effective treatments available as part of an integrative approach to treating IBS and digestive distress. The response rate for chronic gut sufferers with severe symptoms hovers around 80 percent. Plus, the positive effects on IBS symptoms can last for years. This is why Dr. Weisberg routinely uses hypnosis as part of a treatment regimen with the majority of patients he treats for digestive distress.

The advantages of hypnosis are numerous. It is a comfortable and easy way to unleash the healing powers of your mind. Sometimes people also find that hypnosis helps to relieve migraines and tension headaches. If you are surprised at the power of hypnosis as a useful medical tool, it may be because you have been watching too many movies. Let's sort out fact from fiction by examining three sorts of hypnosis:

- Hollywood hypnosis
- Natural hypnosis
- Clinical hypnosis

Hollywood Hypnosis

Hollywood hypnosis is the largely fictional type of hypnosis you see in the movies. When Dracula tells the vulnerable young maiden,

"Look into my eyes," and she loses her willpower and submits to his total control—that is fiction. A myth. When she carries out evil deeds that she would never otherwise do, and when she wakes up and remembers nothing, these too are myths. Hollywood hypnosis is a very useful dramatic device for complicating plots by using mind control to turn the good guys into the bad guys. The only problem is, it ain't real.

It's not just the silver screen that promotes these myths about hypnosis. There are also the stage performers that use Hollywood hypnosis at fairs and other entertainment venues. It is a lot of fun to watch them make a group of volunteers bark like a dog or cluck like a chicken. But that is also a distortion of hypnosis. Dr. Weisberg has used hypnosis in his practice for many years, and while the technique is generally enjoyable for his patients, none of them have ever barked like a dog. Hollywood hypnosis is not going to go away because it has become a time-honored tool in the entertainment industry. But we need to dispel the misconceptions that Hollywood and stage hypnotists perpetuate because they present barriers to hypnosis being accepted by the public as a desirable medical tool.

Common Myths About Hypnosis

Don't be fooled into believing these common myths:

Myth #1: When you snap out of a hypnotic state, you won't remember anything.

While amnesia may occur in very rare cases, people generally remember most of what happened while they were hypnotized.

Myth #2: You can be hypnotized against your will and lose control over your actions.

Despite stories about people being hypnotized without their consent, hypnosis requires voluntary participation on the part of the patient. Numerous studies have proven that individuals cannot be hypnotized into doing something against their will or against their value system. Your clinician does not become Svengali.

Myth #3: Hypnosis will make you lose consciousness.

Hypnosis is not sleep. Ordinarily, you will be conscious of everything that goes on when you are in the hypnotic state. Sometimes, though, you may relax so much under hypnosis that you drift off and lose track of what is happening—or even fall asleep!

Myth #4: Hypnosis will force you to reveal secrets about yourself.

Hypnosis during psychotherapy may sometimes be used to explore unconscious material, but this is done with the mutual consent of the patient and doctor. When hypnosis is used for relief of a physical problem such as IBS or other digestive distress, no such uncovering is usually needed.

Myth #5: People who can be hypnotized tend to be more gullible or simpleminded.

Not at all. In fact, researchers have found that intelligent people are slightly more hypnotizable. It seems that openness to new experiences, rather than gullibility, is related to hypnotic ability.

Natural Hypnosis

It may help you to shake off your misconceptions about hypnosis by realizing that hypnosis is a naturally occurring mental state that everyone has experienced. Researchers have come up with three earmarks of a hypnotic state:

1. It is dissociative. That means you may feel zoned out and have a floating sensation; you become disconnected from the standard perception of time. Think of the last time you attended a boring lecture when time seemed to stand still and your attention to the speaker simply floated away. On the positive side, think of how time flies when you are on a satisfying vacation.

2. It is absorbing. Your attention becomes so focused and concentrated that you become oblivious to the world around you. This is what happens to athletes when they are in the zone. Thousands of people are watching the quarterback when he drops back to make a pass, but he pays no attention to the crowd. Activities such as hobbies are fertile ground for hypnotic states. Think of how time flies and you become totally absorbed in gardening, woodworking, or cooking. Ironically, you are in a similar absorbed state when you watch a movie or a play. You have suspended your sense of disbelief and you forget about the outside world. That's why you hate it so much when someone starts talking in the theater—it breaks the spell.

3. It promotes heightened suggestibility. It is a receptive and open state of mind where the person is likely to accept and respond to suggestions. Once again, while you are watching a movie, you just go along with the action and accept this alternative reality for a couple hours. It is this same openness that makes hypnosis a door to the mind where positive suggestions can enter. This allows for altered perceptions. When your toddler falls and skins her knee and starts crying, you pick her up and kiss the boo-boo away. Suddenly, the pain is much better and she stops crying. Your child just experienced a hypnotic state with heightened receptivity to suggestions. Her perception changed, and it's all perfectly natural.

Clinical Hypnosis

Clinical hypnosis is the normally occurring hypnotic state when a health professional actively uses therapeutic suggestions with patients, usually for the sake of relieving pain or suffering. The hypnotic state is induced by leading patients to focus their awareness inward. That may be the only part of Hollywood hypnosis with any truth in it—though we don't have patients look into our eyes or stare at a shiny pocket watch swinging on a chain.

Standard procedure these days is to guide patients into a hypnotic state with specifically chosen hypnotic suggestions.

Because hypnosis allows people to use more of their potential, learning self-hypnosis is the ultimate act of self-control. Hypnotic communication and suggestions can effectively change certain ways that your body/mind functions. That is why it has proven to be so useful in treating IBS and chronic pain.

Practitioners carry out clinical hypnosis in different ways, but they all encourage the use of imagination. Mental imagery is very powerful, especially in a focused state of attention. For example, a patient with chronic abdominal distress may be asked to imagine what her distressed colon looks like.

Self-hypnosis is the ultimate act of self-control.

If she imagines it as being like a tunnel, with very red, irritated walls that are rough in texture, the patient may be encouraged in hypnosis (and in self-hypnosis) to imagine this image changing to a healthy one, in which the walls of the colon are a light pink color with a smooth texture. This technique often causes beneficial physiological changes.

Another hypnotic method is to present ideas or suggestions to the patient. In a state of concentrated attention, ideas and suggestions that are compatible with the patient's desires seem to have a more powerful impact on the mind. For example, a patient with chronic constipation may be given a suggestion to picture himself in the future feeling increasingly comfortable and confident about having normal bowel function.

What does it actually feel like to experience clinical hypnosis? Dr. Olafur Palsson describes it this way:

> You sit comfortably reclined in an easy chair in a softly lit office. As you listen to the doctor with your eyes closed, you find your body relaxing more and more. Guided by the calm and confident voice, you allow your mind to let go and turn inward. You drowsily notice a mildly curious

floating sensation in your body, as if you are not really sitting in the chair anymore, but rather floating—in the air, or in water. The voice talking to you gradually becomes more distant, and you even find yourself forgetting that it is there . . . but somehow the soothing voice continues to affect you, gently and almost automatically. As you relax even further, your awareness of where you are, why you are there, and who is speaking to you, recedes into the back of your mind. You just content yourself with effortlessly allowing the voice to act on you, and with enjoying this state of profound relaxation and deep calm. . . . You are having a typical hypnotic experience.[4]

How Is Hypnosis Learned and Practiced?

Very commonly, someone will first learn hypnosis by going to a licensed health professional such as a psychologist, physician, nurse, or other licensed clinician who has been trained to use hypnosis in health care settings.

However, Dr. Weisberg reminds his patients that much of hypnosis is self-hypnosis. So technically, when he sees patients for hypnosis consultations, he is basically teaching them how to bring themselves into the hypnotic experience. The goal is for them to learn how to use hypnosis themselves without needing a clinician to get them into hypnosis. The hypnotist often starts by doing some type of induction, which is simply a way to lead the patient into a more focused state of attention, shutting out external distractions. Then, various instructions will be given to deepen the hypnotic involvement, such as counting down or imagining walking down stairs.

Once the patient is in the hypnotic state, the practitioner will give therapeutic suggestions that address the specific problem

---

4 Dr. Olafur Palsson, *www.ibshypnosis.com*, reprinted by permission of Olafur Palsson, PsyD.

the patient is trying to resolve. For example, a suggestion for gut sufferers might be to imagine their gut becoming increasingly comfortable, relaxed, and free from distress with every passing day. Then in the re-alerting phase, when the patient comes out of the hypnosis, the clinician will allow for some type of gradual end of the hypnotic experience, such as counting forward or imagining walking back up the stairs.

You Can Hypnotize Yourself

You don't need to see someone to begin your self-hypnosis regimen. Dr. Weisberg has prepared and recorded a hypnotic experience for helping digestive distress. You can download a recording of this from our website, *www.trustyourgutbook.com*. Allow yourself about fifteen minutes for the experience, and practice it every day as part of an ongoing lifestyle change, just like exercise and proper diet.

Helen's Hypnosis Story

Helen had suffered for many years with constant indigestion, gas, and an unsettled feeling in her lower abdomen. She had gone through so many different treatments with so many different doctors and clinics that she thought she was out of options. Seeing Dr. Weisberg for hypnosis was her last resort.

"I didn't really believe that listening to a recording with someone talking to me about relaxing my body would have any real effect at all. But, being at the end of my rope, I was ready to try even this. As I listened to the hypnosis recording, the strangest thing happened. It's not like I lost consciousness—a part of me could stand back and observe the whole experience as I went through it. But it was like another part of me—my physical body and emotions—were responding to the suggestions very quickly. My arms and legs began to feel quite heavy. As the recording continued, I could always hear the doctor's voice, but it was like the voice drifted further and

further into the background. When I imagined sitting next to a beautiful stream and placed my hand in the cool water, my hand actually began to feel cool and tingly—almost numb. And then, when I put the hand on my abdomen in my mind's eye, I couldn't believe how a strong wave of comfort spread to the exact places where the unsettled feeling bothered me the most! It all passed so quickly that before I knew it, the recording was done. I practice every day, and it has really helped quiet down my unsettled gut."

Breathing Techniques

Breathing exercises are a natural technique for gut relief because, as we mentioned earlier, the gut muscles power the bellows in the breathing operation. The primary role of breathing, of course, is gas exchange: our cells need oxygen, and their waste product, carbon dioxide, needs to be expelled. Breathing is an automatic body function controlled by the respiratory center of the brain. However, we can also deliberately change our rate of breathing, and this is what we do in therapeutic breathing exercises.

Different healing systems from different cultures have long realized the healing benefits of the breath, including yoga, Tai Chi, and some forms of meditation. Many holistic practitioners believe that the breath is the link between the physical body and the ethereal mind, and that spiritual insight is possible through conscious breathing. Regardless of the philosophy, scientific studies have shown that correct breathing can help manage stress and stress-related conditions by soothing and balancing the autonomic nervous system.

Breathing and Stress

The brain sets the breathing rate according to carbon dioxide levels, not oxygen levels. When people are under stress, their breathing pattern changes. Typically, anxious people take small, shallow breaths, using their shoulders rather than their diaphragm to move air in and out of their lungs. This style of breathing empties too much carbon dioxide out of the blood and upsets the body's balance of gases. Shallow over-breathing (hyperventilation) can prolong feelings of anxiety by exacerbating physical symptoms of stress. This can aggravate bloating, cramping, and other symptoms of gut distress.

The Benefits of Achieving Relaxation

When people are relaxed, their breathing is nasal, slow, even, and gentle. Deliberately mimicking a relaxed breathing pattern seems to calm the autonomic nervous system, which governs involuntary bodily functions—including the conditioned response of IBS. Physiological changes of relaxed breathing can provide these benefits:

- Lowered blood pressure and heart rate
- Reduced amounts of stress hormones
- Reduced lactic acid buildup in muscle tissue
- Balanced levels of oxygen and carbon dioxide in the blood
- Improved immune system functioning
- Increased physical energy
- Enhanced feeling of calm and well-being
- A calmed, more comfortable gut

How to Do Basic Abdominal Breathing

There are different breathing techniques to bring about relaxation, but the general aim is to shift from upper chest breathing to abdominal breathing. You will need a quiet, relaxed environment where you won't be disturbed for ten to twenty minutes. Set an alarm if you don't want to lose track of time.

1. Sit up straight comfortably or lie down, and raise your rib cage to expand your chest.

2. Place one hand on your chest and the other on your abdomen. Notice how your upper chest and abdomen are moving while you breathe.

3. Concentrate on your breath and try to breathe in and out gently through the nose. Your upper chest and stomach should be still, allowing the diaphragm to work more efficiently so that eventually you feel your abdomen begin to rise more with each breath while your chest rises relatively less.

4. With each breath, allow any tension in your body to slip away.

5. Once you are breathing slowly and with your abdominals, sit quietly and enjoy the sensation of physical relaxation.

Three Breathing Techniques

In chapter 5, you learned a brief breathing strategy called the 4-4-8 Breathing technique. Hopefully, you are already practicing that simple, effective strategy. Here are three more simple breathing techniques that you can use right away for calming and balancing your mind, body, emotions, nervous system, and digestive tract.

Abdomen-Focused Breathing

Sit in a comfortable chair or recline where you won't be interrupted for the next five to ten minutes.

1. Place a hand on the center of your lower abdomen.

2. Close your eyes and start taking slow, fairly deep breaths.

3. As you inhale, count 1-2-3-4.

4. As you exhale, count 1-2-3-4.

5. To make sure you are doing diaphragmatic breathing, notice the hand resting on your abdomen. If your hand slowly rises when you inhale and falls when you exhale, then you are doing diaphragmatic breathing.

6. Notice what temperature feels most appealing or soothing to your abdomen and use it now in this exercise. If warmth feels more soothing, then on every exhalation, visualize that you are exhaling warmth right through your lower abdomen and through your hand. Alternatively, if coolness feels more appealing, then visualize that you are exhaling coolness through your lower abdomen and right through your hand.

7. Let your attention stay closely focused on the flow of the exhalation through your abdomen and your hand. Do this for five to ten minutes. Not only will your abdomen feel more comfortable, but many people find that the hand resting on the abdomen actually feels warmer or cooler, depending on which temperature they chose.

Sensation-Focused Breathing

People tend to notice their breathing in different parts of their body. In this exercise, we will practice noticing the sensations in three parts of the body: the diaphragm, the chest, and the nostrils.

1. Sit or recline in a comfortable position and start breathing at a normal rate.

2. As you breathe, notice how your diaphragm expands when you inhale and deflates when you exhale. For the next minute, try paying close attention to the physical sensation of breathing in your abdomen as it rises and falls.

3. Next, try feeling the physical sensation of your breath more in your chest. Spend a minute noting how your chest wall expands when you inhale and deflates when you exhale.

4. Now turn your attention to the sensations at the rim of your nostrils as you breathe. Do this for a minute or two. Notice how the breath feels slightly more cool and dry when you inhale, and slightly more warm and moist when you breathe out.

5. Pick the area of your body that was easiest to focus on—the abdomen, chest, or nostrils. For the next five minutes, focus your attention on that spot. During this time, if you should become distracted, simply refocus your attention back on the physical sensation on the place that feels the most natural.

Within five minutes or so, you will feel profoundly calmer and more comfortable, and any gut distress that you may have been experiencing will likely be less prominent and less bothersome.

Complete Breathing

1. Sit up straight with your feet on the floor and exhale.

2. Inhale while relaxing the belly muscles. Feel as though the belly is filling with air.

3. After filling the belly, keep inhaling. Fill up the middle of your chest. Feel your chest and rib cage expand.

4. Hold the breath in for a moment, and then exhale as slowly as possible.

5. As the air slowly goes out, relax your chest and rib cage. Begin to pull your belly in to force out any remaining breath.

6. Close your eyes and concentrate on your breathing.

7. Relax your face and mind.

8. Release any tension that you feel—in either your body or your mind.

9. Repeat, practicing for about five minutes.

Ashley Finds a Bargain

Ashley had spent thousands of dollars over the years trying to find an effective treatment for her chronic intestinal pain and diarrhea. She was taken aback when Dr. Weisberg asked her to do the Sensation-Focused Breathing exercise.

She thought, Are you kidding me? I've spent a ton of money to find help and you're just gonna have me breathe? *But then she thought,* Oh, what the heck? *and started to focus on her breathing following Dr. Weisberg's directions.*

She focused on her breathing and the sensations it caused in her abdomen, chest, and nostrils for about five minutes. Ashley was amazed at the results. "I'm so incredibly relaxed, I feel like an hour passed," *she said.* "Not only that, my abdomen feels soft, quiet, and calmed."

After the exercise, Dr. Weisberg explained the physiology of how it all worked. "I'm a believer now!" *Ashley exclaimed.* "And the great thing is, I can do this for just a few minutes every day, and I get so much out of it."

Meditation

Many forms of meditation have been practiced around the world for thousands of years, usually for spiritual or religious purposes. In recent years, Western science has investigated meditation in the laboratory and has verified its many health benefits, especially its ability to counter stress. Throughout the day, when we experience stress, our body automatically reacts in ways that prepare us to fight or run. In cases of real danger, this physical response is helpful. However, this prolonged state of agitation can cause physical damage to every part of the body. Meditation affects the body in exactly the opposite way that stress does—restoring the body to a calm state, helping it to repair itself, and preventing new damage caused by stress.

Meditation affects the body in exactly the opposite way that stress does.

Health Benefits of Meditation

The benefits of meditation are widespread because this simple act can reverse your stress response, thereby shielding you from the destructive effects of chronic stress. When practicing meditation, your heart rate and breathing slow down, your blood pressure normalizes, you use oxygen more efficiently, and you sweat less. Your adrenal glands produce less of the stress hormone cortisol, and your immune function improves. Your mind also clears, and your creativity increases. People who meditate regularly find it easier to give up life-damaging habits like smoking, drinking, and drugs.

Meditation reduces stress levels and optimizes your immune system, and that has been found useful for reducing symptoms of chronic abdominal pain, bloating, gas, cramping, and indigestion. Recent scientific studies have confirmed that patients who practice meditation achieve better reductions in digestive distress than those who don't. Regular meditation even has the power to activate beneficial genes at the cellular level.

Modes of Meditation

There are dozens of meditation techniques and disciplines available: saying a mantra, staring at a candle flame, counting breaths, and so on. Keep trying various methods until something feels right. And check out community centers, local colleges, and hospital health systems for meditation classes; they're often inexpensive at such places.

Meditation involves sitting in a relaxed position and clearing your mind. You may focus on a sound, like *om,* your own breathing, or nothing at all. It's necessary to have at least five distraction-free minutes to spend. Longer meditation sessions bring greater benefits, but sometimes starting for a briefer time can help you maintain the practice long term. It's helpful to have silence and privacy, but experienced meditators can practice anywhere. Many meditators attach a spiritual component to it, but it can also be a secular exercise as well.

Basic Concentration Meditation Exercise

This concentration meditation method works by concentrating very closely on your breath. Try it as a simple introduction to meditation. You may find it helpful to record these instructions so you can easily listen to them while practicing.

1. Sit in a comfortable chair at a time when you will not be disturbed for ten to twenty minutes. You will focus on your inhalations and exhalations.

2. Just breathe comfortably at your normal rate. As you inhale, silently say to yourself, "So." As you exhale, say, "Hum."

3. For the next ten to fifteen minutes, just continue to breathe and repeat these words internally.

4. After a while, your attention may start to wander off your breath and onto something else, such as, *I wonder what time it is*, or *Did I remember to set up that meeting tomorrow?* or *My back is getting stiff.* This is perfectly normal. It is the conscious mind's normal tendency to wander—the so-called "monkey mind." The goal of this meditative practice is to cultivate concentration. So each time you notice yourself getting distracted, just gently bring your attention back to your breath and to your in-breath and out-breath words. Do this gently, without self-criticism, whenever you get distracted.

As you practice this each day, you will find it easier to focus your concentration on your breath for longer periods of time.

Meditation vs. Hypnosis

Many people think of hypnosis and meditation as similar practices, but there are differences. Both are types of focused awareness that tend to bring a state of calmness and relaxation. However, hypnosis does not always include relaxation, and there are types of hypnosis in which relaxation is not a goal, such as in active-alert hypnosis, which is used by athletes and test takers who need to be focused during their activity.

A primary difference between meditation and hypnosis involves the goal of each practice. In meditation, you focus on a word, your breath, or some other object of awareness without reacting to any of them. The goal is to cultivate quieting and non-reactivity of the mind, which helps lessen stress levels and ultimately calms the body. In hypnosis practice, the goal is often to take advantage of this receptive state to offer specific therapeutic suggestions. For example, "You will find your abdomen feeling calmer and more comfortable with each day," or "You'll find that you can sleep through the night comfortably and without waking."

Both meditation and hypnosis are valuable tools for people with digestive distress, and both are worth learning.

Self-Massage

The sensation of touch is one of the most profound and healing experiences humans share. Often, our mind tends to be completely disconnected from our body because we spend so much time thinking and talking—and so little time feeling. Through touch, we can reconnect our mind and body, allowing relaxation and the unwinding of stress and tension. Massage, like meditation, is an ancient practice that comes in many forms. Most types of massage aim to release tension from the muscles and connective tissues.

Digestive distress is closely related to the function of muscles and connective tissue in the digestive system. In irritable bowel syndrome, the smooth muscle tissue that lines the walls of the intestine contracts more strongly than normal, which forces food through the colon faster. That's one of the reasons we get diarrhea and gas. Sometimes the intestinal muscles contract in such a way that food passes through too slowly, and constipation follows. And because stress throughout our

Digestive distress is closely related to the function of muscles and connective tissue in the digestive system.

body contributes to intestinal distress, the release of that tension is certainly beneficial for gut sufferers.

You may not always be prepared or able to receive a healing touch from a trained massage therapist. However, learning to reconnect with your body by yourself can be a profound practice. So let's learn a few self-massage techniques that you can use to improve your health.

Morning and Evening Circulation Massage

Do this exercise every morning and evening. It helps strengthen the body, stimulate blood circulation, and relax nerve endings.

1. Lie comfortably on your back. Using your fists, gently thump the outside of your body, starting with your legs and arms, working from top to bottom.

2. Then move inward to your torso and thump from bottom to top.

In the morning, this self-massage technique will waken and prepare your body—and mind—for the day ahead. Before bed, this massage calms the mind and drums out the stress and tension of the day.

One warning: If you're taking any kind of blood thinner, such as Coumadin (warfarin), check with your health care practitioner before beginning this practice, as it might cause bruising. Also, as with any other exercise suggested here, if it feels uncomfortable to you, feel free to discontinue it.

Post-Meal Belly Rub

Most of us rub our belly after eating instinctively, especially after overeating, but if you're not in the habit, this technique can help alleviate digestive distress.

1. Place one or both palms on your abdomen.

2. Rub in clockwise circles. This is the same direction food naturally moves through your intestine, so your circular massage will help to stimulate digestion.

This is particularly helpful for people who have constipation-predominant symptoms.

Mild Abdominal Massage

This simple exercise does wonders for relieving tight muscles.

1. Lie down on a comfortable surface. Take two or three minutes to settle in and breathe comfortably.

2. Feel around your abdomen for any areas that feel particularly sore or tight.

3. Take a deep breath, and as you exhale, gently press and massage the sore area. If this feels comfortable, you can continue to massage any particular area through a few breaths.

4. Repeat this for all sore and tight areas, remembering to exhale as you gently massage.

Natalie Tunes in with Her Gut

Natalie was a fifty-two-year-old music teacher who suffered with constipation for several years. We introduced self-massage as a way for her to alleviate the abdominal tightness that accompanied her chronic constipation.

"It's funny," she said. "I've had digestive distress for so many years, and yet it never occurred to me to bring the power of touch to this problem. I know we all rub our tummy sometimes when we have indigestion, but I never realized before just how tight the muscles feel in my lower abdomen and stomach."

Natalie began practicing self-massage and was amazed how loose and comfortable her abdomen felt afterward, even when her muscles started out as tight as a piano string.

"Every time I do self-massage, it brings me a feeling of circulation and calmness," said Natalie. "It really calms me down. I look forward to doing the self-massage every day now. It's nice to know that there's something I can do that makes a difference in helping me feel better."

Your Daily Dose

It's helpful to have a variety of calming and balancing techniques in your repertoire so you can call on them when you need them. Some people feel more attracted to the self-massage techniques, while others prefer self-hypnosis or meditation practice. Most everyone takes advantage of the breathing techniques, as they are brief and very easy to use.

We recommend that you experiment with various methods that we showed you in this chapter, find a few that you like, and commit to practicing them every day. Your daily dose of calming and balancing will provide you with a foundation of good health. As you practice these methods more often, they will help your body/mind to be more receptive to the central healing tools of this book: Neurohormonal Retraining and Ecological Rebalancing.

11

Resolve Difficult Emotions and Their Physical Effects

The organs weep the tears that the eyes refuse to shed.

—Proverb

E motions present a great dilemma in any society, and hence in any individual. If emotions are given free rein, external chaos erupts. Yet if they are kept strictly under control, the repressed emotions can cause internal chaos. This is especially troublesome for gut sufferers because unresolved difficult emotions are one of the most common triggers for flare-ups of bloating, cramping, abdominal pain, gas, and other gastric distress.

The ill effects of this repression are among the major causes of a relapse for those who have made significant progress in the CORE program of reducing intestinal distress. That's the purpose of this chapter, to provide skills to resolve difficult, negative emotions, and their physical effects. These problematic emotions include sadness, anger, shame, disgust, anxiety, frustration, grief, and despair.

The Shame Barrier

Emotions evolved in humans because they were essential for the survival of the species. For example, the emotions of fear and anger are essential for avoiding harm at the hands of an aggressor. The emotion of love is essential for bonding with people. If prehistoric people did not bond with their babies and protect them, the infants could never have survived on their own. Game over. The emotion of disgust may have evolved initially as food aversion to avoid eating things that could make you sick or even kill you. As societies developed, each culture faced the challenge of how to control people's emotions.

In Western culture, reason was anointed to rule over the emotions, and rationality became the idealized state of mind. Emotions were irrational and not to be trusted. That's why so many people are embarrassed or ashamed to admit they have emotional problems. There is a sense that emotional problems are not real, so the person who has them is often considered to be weak, mentally ill, or just faking it.

It's no surprise that many gut sufferers feel limited in how to handle their difficult emotions. Often they try to avoid addressing their feelings because of the perceived stigma of being depressed or anxious. Some gut sufferers believe that if they acknowledge their negative feelings, this confirms that their digestive distress is psychosomatic, and therefore not real. For many people with chronic intestinal distress, this is a charged topic. Bring up sadness, anger, shame, or fear with them and you're likely to get a scowl or defensive reaction: "This isn't psychosomatic! I have a real physical problem!"

We agree with you—it is not psychosomatic! Your digestive distress is not all in your head! Our brain and body do not distinguish between whether something is physical or psychological. After all, it's not physical, and it's not psychological: it's psychophysiological. The point is that any given stimulus can activate both the bodily systems and the emotions, and what matters is that the gut gets imbalanced, dysregulated, and oversensitized.

Reticent Rhonda

Rhonda was a twenty-one-year-old college senior with mild to moderate nervous stomach, as she called it. The pain was only moderate, but she frequently felt shaky and experienced rapid onset of gut distress throughout her lower abdomen.

Dr. Weisberg's assessment revealed that she didn't have any particular food triggers for her symptoms and she ate a fairly balanced diet. So he asked her, "Can you think of anything else that might be causing trouble in your gut?"

Rhonda hesitated for a moment and said, "I tend to worry a lot about all kinds of things: my grades, my boyfriend, whether I'm going to find a summer job, and whether my stomach is going to get worse. But I don't want to spend time talking about this because it's not a big deal, and I can handle it okay. I can handle my worries and anxiety just fine, and I'd rather not talk about it anymore. Anything else you can do to help me?"

Dr. Weisberg suggested that she start by keeping a symptom diary. She should write down when she had symptoms and their severity, and also what foods, external events, and emotions/thoughts were occurring around the same time.

Difficult Emotions

The relationship between difficult emotions and gut distress is a bit like the chicken and egg conundrum. Emotions can make gut distress worse, and the distress can make the emotions worse. Strong emotions can sensitize the nervous system, which in turn causes pain, bloating, and cramping in the gut. At the same time, the pain itself can cause feelings of frustration, panic, or anger. When anxiety leads to increased worry, this can make you more hyper-focused on minor fluctuations or spasms in your abdomen, which accelerates the vicious cycle of gut distress.

We are all hardwired to experience emotions in response to life's events.

Emotions are not just something that weak, passive, or neurotic people experience. We are all hardwired to experience emotions in response to life's events. It's important to realize that all emotions, especially intense negative emotions, are *biological events* that get translated into the language of the body. This can set off cascades of neurotransmitters and hormones that directly communicate with muscles, connective tissues, and—of course—with your digestive system.

You know by now that difficult emotions don't *cause* IBS, but they can have a profound effect on the course of your gut symptoms. Many clinical studies tell us that patients with IBS have stronger emotional responses to uncomfortable sensations in their abdomen than non-IBS sufferers. This is why Neurohormonal Retraining is so important for healing.

In our clinical practice, every day we see patients whose cramping, bloating, and pain have been made worse by difficult emotions. Nevertheless, for many, their feelings remain a taboo topic until they reach a point of such intense gut distress that they can't stand it anymore.

The Clue in Rhonda's Diary

At her second appointment, Rhonda brought in her symptom diary. As she reviewed it with Dr. Weisberg, it was clear that her unstable stomach episodes corresponded to times she worried about school, work, or relationships.

With great hesitation, she asked, "Doctor, does this mean that my gut problems are because I'm crazy and neurotic?"

Dr. Weisberg reassured her that she wasn't the least bit crazy. He went on to discuss how tough emotions like anxiety can communicate directly to the brain, the gut, and the nervous system, making her shakiness and instability in the abdomen feel intensified. "The good news," Dr. Weisberg said, "is that this is very common and very treatable. Everyone gets anxiety and worry from time to time. Our goal is for you to learn new ways to respond to the anxiety and worry so you don't fight it so much. This will help your gut feel better."

Rhonda breathed a sigh of relief. Then she smiled and said, "Let's get started."

It's not the difficult emotion itself that causes the physiological disruption. Rather, it's how we handle and respond to the feelings that make the biggest difference in how much physiological disruption occurs.

Why Can't We Just Ignore or Push Past Negative Emotions?

Neurologist Robert Scaer brought up an interesting question at a conference: Animals in the wild are under constant stress every day, because other animals want to kill them and eat them. Every day is a struggle for survival. Why don't animals in the wild develop post-traumatic stress disorder and other emotional dysfunctions?

The answer: animals have natural, instinctive responses for discharging negative emotions. Dr. Scaer showed a video of researchers tracking down polar bears in the Arctic and shooting them with tranquilizer darts. This does not physically harm the bear, but it certainly traumatizes it. When this huge polar bear awoke from the tranquilizer around thirty minutes later, it first laid motionless. Researchers quickly took physiological measurements of the bear during this time and recognized that it was in what they called a "freeze response." This is the deer-in-the-headlights syndrome, an innate and automatic state of immobility in response to certain threats. After the bear laid awake in this frozen position for about an hour, it suddenly started trembling and shaking uncontrollably. Its torso, its head, its limbs shook continually for about forty-five minutes. When the trembling ended, the bear simply got up on all four limbs and calmly walked away as if nothing ever happened! There was no unresolved trauma left to cause him any future trouble.

For humans, however, the issue of resolving difficult emotions is more complex. Unlike animals, we *Homo sapiens* have a large cerebral cortex, the cap that covers the top of the brain. This evolutionary advantage allows us to conceptualize and think abstractly about the past and the future and to rationalize our actions. It also allows reason to suppress emotions. We don't just shake it off like the bear did.

We humans can think things like, *Oh, I shouldn't feel so angry at my mother just because she said I wasn't dressed nicely enough,* or *I'm not going to let myself cry in front of all these people—they'll think I'm weird,* or *I'm so upset that I feel like trembling—but there's no way I'm going to do that in front of my family. They'll freak out.*

Emotional Expression

Think of infants and toddlers. One moment they're crying, and the next moment they're laughing. Suddenly they're irritable, and the next moment, they're filled with delight and curiosity. As new emotions emerge from moment to moment, they are expressed. Have you ever seen a depressed toddler? Probably not, because they have not yet learned how to suppress their emotions. That only happens when they get older and become socialized and self-conscious.

All of us occasionally try to ignore or muscle through tough negative feelings. Maybe you're in a difficult work environment, and it isn't the time or place to express your frustration with an overcritical boss. You could be annoyed at your significant other, who doesn't want to hear how you feel ignored in the relationship. Perhaps you're sitting at home alone, making way too many trips to the bathroom, and it seems like nobody in the world can possibly understand the pain you're going through. It's very tempting at these times to try to ignore your feelings of loneliness, despair, anxiety, and bitterness. Sometimes you don't want to have to deal with your emotional pain because you're too busy. You have a last-minute assignment to complete or your daughter has to get to the soccer game.

The problem is that it's not just an occasional thing. People have learned to ignore their negative feelings *most* of the time, not just when they're busy or distracted. That's when chronic difficult emotions start to wreak havoc on the body. While we're intellectually absorbed with our rational cerebral cortex, our primitive limbic brain—sometimes known as the reptilian brain—bypasses rationality and continues to send signals out to the entire physiology.

Bill Denies His Nervousness

Bill was sitting nervously in the waiting room, wondering how soon he would be called in to the doctor's office. He was being examined yet again for his chronic cramping and bloating pain. He kept looking at the clock and glancing at the receptionist while trying to read a magazine.

He noticed his palms were sweaty when he saw a Sports Illustrated, *but he tried to forget it. As he reached over to grab the magazine, he felt his shoulder stiffening up, but pushed the sensation out of his mind, thinking,* What's happening in the NBA playoff? I gotta read this. *He could feel his anxiety mounting, he could hear his heart pounding, but he didn't want to think about it.*

So he sat there and stared as hard as he could at the magazine article. As he tried to focus on the words on the page, he could feel his legs getting fidgety and weak. After what seemed like hours, the clinic staff finally called his name to see the doctor. He looked at his watch as he stood up—he had actually waited only ten minutes.

The moral of this story is that difficult emotions can seriously alter your digestion, even if you try to avoid or ignore your feelings. It's human nature to ignore anything unpleasant, whether it's physical pain or emotional pain. It's all the same to our pain-processing center. But these difficult emotions need to be identified and addressed so that their disruptive effect on your gut can be reduced. Otherwise, to quote a wise colleague, it becomes an example of "secrets kept from the mind but not from the body."

When we have a difficult or painful experience and we can't cope with it out of fear or pure avoidance, we sometimes just dismiss the emotion. We get busy, we exercise more, we have a drink or eat a bit more, or we try to pretend it didn't happen. When we have successfully buried the emotion, we're not consciously aware of it anymore. But the trick is on us—and in us. These feelings stay embedded in our deeper brain structure as well as in our muscles, ligaments, stomach, colon, and connective tissue. They're buried but not dead. Outside of our mind's eye, these repressed feelings are altering our brain chemicals. The longer these emotions remain suppressed within the chronic gut sufferer, the more damage they can do. It is only when we embrace our emotions and feel them that they quit haunting us.

How Do I Suppress Thee? Let Me Count the Ways

The following are a few examples of the patterns people establish to avoid experiencing their emotions:

- Ignoring feelings
- Pretending something hasn't happened
- Overeating
- Eating foods loaded with sugar and fat
- Eating when not hungry
- Excessively drinking alcohol
- Using recreational drugs
- Overusing prescription drugs, such as tranquilizers

- Exercising compulsively

- Engaging in any type of compulsive behavior

- Excessive sex with or without a partner

- Always keeping busy

- Constantly intellectualizing and analyzing

- Excessively reading or watching TV

- Working excessively

- Keeping conversations superficial

- Burying angry emotions under the appearance of calmness

Notice that many of these strategies involve using food in a way that can aggravate your digestion. This becomes a double risk factor for gut sufferers: both the difficult emotion that is ignored and the eating pattern used to push it away can worsen IBS symptoms

The CORE method emphasizes benefiting from the wisdom of the body, whether it's a gnawing sensation in the lower abdomen or a pounding chest filled with anger or anxiety. Quite often, a negative emotion is sending you a signal in an attempt to get you to pay attention to something. It may be trying to guide you in a new direction, as if it were an internal GPS. It could be trying to awaken important feelings regarding your work, a loved one, a friend, or an action you need to take.

Recognizing and dealing with negative emotions is a vital life skill. It is a strength. It not only promotes vibrant health and reduces flare-ups of digestive distress, but it's also crucial for preventing depression and anxiety as well as reducing suicidal thoughts and impulses.

Recognizing and dealing with negative emotions is a vital life skill.

Often when gut sufferers see a physician about depression, anxiety, or panic, they're advised to control the feelings with antidepressant medications. A review of several blogs on the topic of IBS and emotional distress reveals that many gut sufferers agree that the primary option for responding to emotional distress is antidepressant medication.

We don't agree. The CORE program is based on an integrative medicine orientation. We recognize the value of occasional use of antidepressants as a bridge to help someone stuck in persistent feelings of severe depression or anxiety. We see antidepressants as one of many tools to help reestablish psychophysiological balance and equilibrium, but certainly not the primary one. Antidepressants are a short-term solution that should give way to longer-term, sustainable self-care treatments. In the long run, skills are more effective than pills, and there are skills that can manage and reduce the disruptive effects of negative emotions on your digestion.

Dealing with Difficult Emotions: Myth and Fact

The pressure cooker model is an outdated paradigm for understanding emotions. This model supposes that suppressed feelings keep piling up and building more pressure until they finally explode. The mode of emotional release is volcanic—like a tantrum or an emotional breakdown. Conversely, it could implode into depression or a physical illness. A variation is that difficult repressed emotions can be resolved by yelling, screaming, beating on a pillow, or confronting someone head on. Again, this model is obsolete.

Interestingly, you don't have to do *any* of these things to tame the ill effects of unresolved negative emotions. Ultimately, the solution comes down to our old friend self-awareness: being willing and able to feel and recognize the emotion in your body. All you have to do is recognize and express. Let's look at each of these steps more closely.

The Two-Step System to Resolve Negative Emotions

These two steps for successfully resolving difficult negative emotions sound deceptively simple but are very effective when applied regularly:

1. Recognize: Acknowledge and feel the difficult emotion within yourself.

2. Express: Choose from a continuum of possible constructive actions to release the buried emotion.

Recognize

Recognition of an emotion means "identifying the feeling in your body and acknowledging to yourself that the feeling is there." Healing can occur when you allow yourself to feel, express, and release emotions and unmet needs from the past that you have suppressed or tried to forget.

Behind every negative emotion, there is an unmet or unresolved need, either from the present or the past. Connecting with your feelings is to become aware of your deepest needs and desires, and to get in touch with the wisdom of your inner-knowing resources. Just as important, by recognizing your difficult emotions, you are on your way to reducing a major trigger for further flare-ups of gut distress.

Josh's Secrets Kept from the Mind but Not from the Body

Josh, a twenty-six-year-old barista, suffered from chronic gut distress. On Thanksgiving, he went to have dinner with his parents and his three siblings. In Josh's family, there has been an unspoken rule that no unpleasantness should be brought up. He was trained to ignore any smoldering conflicts with his siblings or his parents. Josh has always tried to look on the bright side.

Accordingly, Thanksgiving dinner was outwardly pleasant, reserved, and quiet. No controversial topics were discussed. However, at one point his father pushed one of Josh's hot buttons. He said, "Josh, isn't it time you find a job that makes more money?"

Josh had told his father before that this was a sensitive topic and that he would rather his father didn't bring it up again until Josh was ready to talk about it. All Josh said now was, "Yeah, you're right, Dad."

The evening went on with the traditional veneer of calm and polite conversation. But internally, Josh was churning, both emotionally and physically. By the next morning, he had severe diarrhea and painful abdominal cramps. He tried to tell himself he wasn't bothered by his father's comments, but his gut knew the truth.

When you experience an emotion, ask this question: what do I feel? Be honest with your feelings and name your experience as best you can. Allow yourself to feel fully and completely. Observe the thoughts that go through your mind and the sensations that you feel in your body. Don't judge. Just watch, listen, and feel.

Don't Ask Why
When you notice a difficult emotion, do not ask *why* the feeling is there. The act of raising that question activates different parts of the brain than the ones you use to recognize the feeling. Just stay with the experience unquestioningly. The reasons *why* the feeling is there will make themselves clear before long. As we have been emphasizing throughout this book, you can't change it if you can't feel it. Awareness is the first step of change.

Methods to Identify Your Emotions

Emotions are reliable indicators of what is really going on inside of you. There are many ways to identify emotions, so explore which methods best fit your personality. Some people need solitude, whereas others need to be around other people to best get in touch with this. Some want to write everything down, while others use a much more casual approach. Sometimes a combination approach is best for identifying deeper emotions.

What follows are several methods to identify what you are feeling about a person, place, situation, or thing. Find the ones that suit you and use them to help you on your journey toward emotional recognition.

- Neurohormonal Retraining: This powerful method for reducing digestive distress also works beautifully for identifying your emotions. The first step is to stop and recognize that you may be feeling something. This may start as a hunch or inkling that something is stirred up. Or someone may have said something that, as you think about it, seems like it might be insulting or hurtful to you or someone else. Once you recognize that you may feel a little bit of sadness, anger, or anxiety, scan your body to notice where you can tangibly sense the feeling. Remember—feeling refers to both emotions and bodily sensations.

- Find the part of your body where you most notice the trace of the emotion. It may be a slight tightness in the chest. It might be a subtle rumbling in your lower abdomen. It could be the sensation of your neck and shoulders tightening up. It might be a tingling sensation in one or both arms. Wherever it is, use the Neurohormonal Retraining procedure. Focus on the sensation very closely without trying to fight it, analyze it, or make it relax. Just imagine that you're going to pull a chair up right next to it and pay very close attention to the feeling exactly the way it is. If it moves and changes, just stay with it and make room for it to be there the way it is. This is a very effective way to listen to your body and gain awareness of what emotions may be stirred up.

- Listen to your thoughts and daydreams: We become so accustomed to thinking in certain patterns that we are no longer aware or conscious about our thoughts and daydreams. Catch those daydreams, hold the thoughts, and bring them up into your conscious mind. This will tell you a great deal about yourself, your relationships, and what you love and hate. Try to keep a written diary of these passing thoughts and daydreams. Writing them down will help you to understand your thought patterns and bring them into a higher level of awareness within you. Over time, important patterns and clues about your emotional experience will emerge from your written record.

- Identify your little and unimportant hurts: Many people walk around saying, "It's not important," or "It doesn't matter" when in fact a big piece of hurting emotion is buried within them. It *does* matter. Men in particular are socialized to frequently underplay what bothers them. Instead of sweeping these little hurts under the rug, write down a detailed description of all the small bothers that won't go away. Put on the list every little hurt that you keep remembering, regardless of when it happened. Many people have a bunch of these little hurts inside, lingering since childhood. These buried emotions create difficulties with your health. Identifying these hurts will tell you a great deal about your frozen emotions.

- Write in a journal about anything that prompts an emotional reaction: Keep an ongoing record of any noticeable emotions. Add the cause to your list, whatever it is—the weather, the traffic, your spouse, children, politicians, the stock market, your boss, other people. Try to identify what really made you angry. Again, don't ask *why* at this point. Writing these details down, even briefly, will help you see contributing factors much more clearly, increase your awareness, and help you to know your emotional self at a much deeper level.

Gwen Gets the Feeling: Part One

Gwen had spent the evening with her boyfriend, Nate. He seemed somewhat distant and cold to her, but whenever she asked him about it, he said everything was fine. They went to dinner and to a movie. At the end of the evening, he dropped her off at her home, and then he headed home.

As Gwen sat in her bedroom later that evening, she felt a gnawing in her lower gut. She pulled out her journal and wrote: I don't know what to do about it yet, but I know that it felt awkward and bad being with Nate this evening. Something is wrong, and I feel hurt and rejected by him.

- Be aware of memories that won't go away: If you repeatedly remember certain situations or hurts that happened some time ago, you are guaranteed to have unidentified or unresolved emotions around this person or situation. You will need to pull this situation out and experience it, recognizing the hurt around it. Try to document these awarenesses carefully, as they are quite likely to be causing you much physical distress. Forgiveness is something that occurs as a result of owning and releasing your emotions. We often reach for forgiveness without doing the work required to release emotions of hurt and anger. Forgiveness is a result of an emotional process. There are no shortcuts.

- Be specific: People get confused when they are trying to get to know their emotions because they speak in general terms rather than pinpointing specific emotions. A good example of this is depression. If you are depressed, you may be experiencing loneliness for people, loneliness for God (spiritual loneliness), boredom, or a lack of creativity in your life. You may be feeling abandoned because of a death or divorce. Or, you may be angry and then suppress the anger, which gets experienced as depression. If you just say you are depressed, you will have great difficulty releasing the emotion or finding a solution to the situation causing the emotion.

- If you have difficulty finding words for the feelings that are emerging, refer to *www.trustyourgutbook.com* for a list of feeling words. You're sure to find a term that aptly fits your experience.

- Be aware of compulsive or excessive behaviors: A good clue that you are ignoring important emotions is if your behavior suddenly changes. You may go for weeks or years acting in a manner that seems normal, but then you may find yourself overeating, working excessively, drinking daily, engaging in compulsive sex, or many other types of compulsive behavior. Excessive behavior such as exercising or gambling too much can overwhelm your feelings. You may be acting this way because you are afraid of these feelings and where they might lead in your thinking and actions.

A Method for Uncovering Hidden Emotions

Here's a useful method to dig for underlying emotions. If you find yourself about to engage in another round of a compulsive behavior, whether it's grabbing another handful of cookies late at night, pouring one more glass of wine, or something else entirely, do the following:

1. Stop what you were about to do.

2. Write down this question in your journal: What would I be feeling right now if I did not _____ (list compulsive behavior here)?

3. Write the answer.

By doing this, you are taking advantage of a valuable opportunity to identify the negative feeling that is being suppressed by the compulsive behavior.

Try to identify the times when your excessive behavior was triggered and, as soon as you can, identify the emotion beneath that behavior. It can be stress or fear related to a new job, the death of a friend or partner, difficulties with lovers or children, or stress and anxiety about your chronic digestive pain and cramping. Document these emotions as best as you can.

- Notice positive emotions: The way our emotions are hardwired, you cannot just suppress one emotion. When you limit or repress one emotion, you end up suppressing *all* of your emotions. Therefore, by recognizing any positive emotions you experience, you also help to open the pathways to recognizing negative emotions as well. There are additional reasons for you to identify your positive emotions during these exercises. You may well be feeling love, care, compassion, trust, forgiveness, and generosity, perhaps even many times in a day. Noting this provides a more realistic and well-rounded self-awareness. If you record only negative emotions, your picture of yourself will be quite distorted and lacking in balance. You have the capacity for a full emotional experience, and every emotion needs to be recognized for healing to occur.

By recognizing what negative emotion you are experiencing and acknowledging it to yourself, you have already taken the first and most important step in reducing the intensity and disruptiveness of this feeling on your gut. Congratulate yourself for starting to use these methods. As you practice, you will find it easier to quickly and effectively identify and acknowledge your difficult feelings.

Express

After you have identified the emotion within you, then you must decide if some type of action is necessary to express the emotion. To maximize your ability to effectively resolve negative emotions, it will be helpful to expand your repertoire of ways to express emotions.

Expression of important emotions enhances a more balanced autonomic nervous system, healthier immune functioning and generally improved physical health. When you are experiencing strong feelings—anger, dread, sadness, or anxiety—recognizing and expressing the feeling can reduce the likelihood of a painful abdominal flare-up. Choose from a continuum of possible action responses for expressing the emotion.

Suppose you are angry at a friend for saying something unkind to you a party. Imagine a continuum of possible levels of emotional expression. At one end of the continuum would be doing absolutely nothing, not saying a word. At the other end of the continuum would be confronting the friend directly and angrily about what he said. In between lies an unlimited range of possible responses. The more options you have, the more flexibly you can respond to the unique details of each situation.

Here are some possible options for expressing your emotions:

- Recognize: Acknowledge the emotion and admit to yourself that you feel angry about what took place. Sometimes this is all that is needed.

- Express: Talk about the feeling to a different friend or confidant. Sometimes talking about the emotion and having it witnessed leads to a greater impact and feeling of resolution.

- Walk and talk: Sometimes people feel stuck when immersed in a strong, difficult emotion. Getting up and moving your body can help to dislodge the challenging feelings. A variation on this is to go for a walk with a voice recorder, so you can talk spontaneously on your stroll. Many people find that this helps facilitate effective emotional expression.

- Write in a journal: Write about the emotions stirred up by the event. Make sure that, as you write, you use short, simple phrases and include feeling words. An example of this would be: *When I heard you tell Joni at the party that I seem to be getting sick so much lately, it made me feel angry and embarrassed.*

- Write an unsent letter: Writing a letter and then not sending it can be a very helpful strategy for expressing difficult emotions. You can feel free and uninhibited to fully express the extent of the anger, hurt, embarrassment, or frustration because you know the letter will not be sent to the person you are writing to. An added bonus: this technique is especially useful when the person you have strong negative feelings toward is deceased.

- Use the empty chair technique: In this variation on the unsent letter, sit down and face an empty chair, imagining that the person you want to address is sitting in it. Take a moment to clearly imagine that other person in the chair. Then, fully express all your thoughts and emotions to the person in the empty chair. Although your friend isn't actually there, you'll still get the same emotional benefits, because you will have done the necessary steps of recognizing and expressing the difficult feelings.

- Confront the person: You can talk to the person you feel angry at, but write an unsent letter to him first to crystallize and clarify your emotions before speaking with him. Sometimes nothing substitutes for talking directly to the person who is bothering you. By writing out your thoughts and emotions first, you are more likely to successfully convey all that you need say for constructive resolution of your negative emotions.

Gwen Gets the Feeling: Part Two

Gwen decided to write an unsent letter to Nate. In it, she talked about how much she cared for him and how much she wanted to grow closer to him. She wrote, Even though you told me nothing was wrong, I felt hurt and rejected by how you were treating me the other night. I want us to have better communication if our relationship is going to thrive. *A few days later, she spoke with him directly about these feelings; Nate said he would think over what Gwen told him. Even though the problem with Nate wasn't resolved, Gwen felt better because she was no longer suppressing her emotions. The gnawing in her gut had disappeared.*

As a general consideration, remember that calm, clear identification and expression of your difficult feelings is more important and effective than yelling and accusations. Don't be a throwback to the pressure cooker school of thought. Use the format of making "I feel" statements. For example, rather than saying, "You're so mean and insensitive for what you said," it's much more effective to say, "I feel angry and embarrassed when you talk about my health to others like you did with Joni." Remember also that expression of difficult emotions can be achieved without necessarily even talking directly to the person whom you are having the reactions toward.

These are just some of the many possible strategies for expressing and resolving negative emotions. Practice several of them, and talk with your family and friends about what methods they use. Good management of negative emotions is one of the best medicines you can use for reducing IBS flare-ups and improving your long-term abdominal health.

PART IV

Ensure

Change is never easy without centering your body/mind, observing for awareness, and restoring core strengths. Lasting change requires additional special considerations. Our wish for you is continual discovery, development, and strengthening of your capacities to ensure harmony, balance, and resilience.

Over the years, our patients have taught us a great deal about the challenges encountered when somebody aims to make a lasting change in their self-care lifestyle. We have shared these journeys with them, and in part IV, we share with you what we have found to be important and helpful in maintaining your new skills so that they become healthy habits and lifestyle changes that last a lifetime.

12

Maintain Your Gains and Achieve Sustainable Lifestyle Change

Motivation is what gets you started. Habit is what keeps you going.

—Jim Ryun, track athlete and author

Part of the philosophy of a holistic approach to the human condition is that life is not made of straight lines; life is made of circles. The biggest difference between Western philosophy and Eastern philosophy is that the West values things that are unchanging and eternal. That's why they fell in love with numbers and trust a quantitative view of life more than a qualitative one. In contrast, the East embraces change. The great Chinese book on how to handle life, the *I Ching*, literally means "The Book of Changes." The great yin yang symbol is a circle in which the two great forces of the world constantly rotate in perpetual change. Buddha taught that one must accept impermanence as a basic principle of existence; all is in flux.

The Western view of action is a straight line, with the end of the line as a goal. It is the schema for all drama: everything has a beginning, a middle, and an end. Ingrained in Westerners is the fairy tale format we were raised on. Once the beautiful young maiden finds her Prince Charming, they live happily ever after. The end. This is not the end.

But this book is not a fairy tale. We feel more akin to the Book of Changes, which reminds us that we need to become accustomed to change. Often, when people feel better, they wish and hope that they will never have a setback and that they will not have to keep paying attention to the practices that help their condition improve. We encourage you to make friends with change. You are constantly being challenged to reestablish your balance by the changes in the world and in yourself. This is not pessimistic; this can be hopeful, and even empowering, once you're aware of it.

The skills you have learned from this book can enable you to rebalance your life and restore your trust in your gut as your center. This balance restores the health of your body/mind system and relieves your distressful symptoms. But there is one more important set of skills to learn: the ability to maintain your balance despite the inevitable cycle of change. We also want you to continually practice the skills you learn in the CORE program so they become ingrained habits. Make sure that what you have been working hard to achieve becomes a sustainable lifestyle change, not just a temporary improvement.

Ensure Your Perpetual Balance

It is all too common for people to resolve to change how they live their lives, only to run out of steam in a few weeks and slide back into the old habits. We are going to show you a number of tools to make sure that you stay on track and tips to ensure that the changes you make will become long lasting.

One of the lasting lessons we have learned is that lifestyle change is complex. It is not simple—it takes plenty of twists and turns. Over the years, our patients have taught us a great deal about the challenges encountered when somebody aims to make a lasting change in his or her self-care lifestyle. We want to share with you important and helpful ways to maintain these new skills so they become healthy habits that last a lifetime.

Maintaining New Self-Care Behaviors Can Be Challenging

In your personalized CORE program, we're asking you to take a number of steps to empower yourself: make different dietary and nutritional choices; learn to quiet and balance your system; and refocus your attention to retrain your brain, gut, and nervous system. All of this requires practicing new skills. One of the best ways to keep your healthy changes more permanent and on track is to anticipate what obstacles may interfere with your new habits. By successfully navigating through those pitfalls, you can make these important changes part of your daily routine.

Every New Year, about half of all Americans make a resolution about how they are going to improve their lives. They swear they will lose weight, stop smoking, exercise regularly, or make other longstanding lifestyle changes. Researchers have looked at the success rates of these resolutions: the first two weeks usually go along beautifully; by February, people are backsliding; and by the following December, most people are back where they started, often even further behind. What happened?

It is certainly not surprising because we all know people like this. Even when they have health problems and their physician commands them to make lifestyle changes, most just quit trying after a while and just sputter out. A smoker with a three-pack-a-day habit quits for several months, only to start up again in less than a year. A woman goes on a diet and loses twenty-five pounds, only to regain it more quickly than she lost it. A man gets bariatric surgery to restrict the size of his stomach, but he gains weight anyway and then starts drinking out of frustration. Why do so many people not keep their resolutions? Are people just weak willed or lazy?

No. It's not a matter of willpower. Most people are not lazy; they genuinely want the change they seek. The problem is that the rest of their life does not accommodate their new habit, so it cannot get rooted. You need support for your new habits because they do not exist in a vacuum. The holistic approach is to look beyond the new skill to other areas of your life and your mind, to other people, and into the future. This broad attitude of finding the larger context gives you a substantial advantage over your average New Year's resolution maker.

Without techniques and skills to assist them in overcoming the unforeseen challenges of lifestyle change, people may tend to rationalize their failures. Many are used to their daily routines and find change to be foreign, awkward, and uncomfortable. Many people make resolutions for change that are unrealistic or out of alignment with their real feelings about themselves. Sometimes their level of distress just isn't severe enough to make it worth their while to dedicate themselves to the ongoing commitment of changing their daily habits. Sometimes the change just takes too much time or energy.

But you can rise above these excuses and overcome your gut distress. To counter the natural inertia that fights change, we will present some support skills to ensure that your new self-healing habits can grow with you for the rest of your life.

You can counter the natural inertia that fights change.

The Three Daily Steps

These daily activities can ensure your continued success with the CORE program:

- Each morning, set your intention.

- Each day, wear or carry a physical reminder of your intention.

- Each evening, express gratitude for every positive step you were able to take.

The Five Weekly Reflections

These weekly reflections will ensure continued progress with the CORE program:

- I have made progress in these ways . . .

- I have noticed these positive changes . . .

- I have struggled with . . .

- I could use help with . . .

- Today, I would like to celebrate . . .

Supportive Skills to Ensure Sustainable Change

These skills will aid you in solidifying your adherence to the CORE program as a permanent lifestyle change.

Build on Your Strengths

Whenever you are striving to make changes in your daily life, it is important to identify and use your innate strengths, talents, and skills. If you're like most people, you have a number of skills and resources available to help you make those changes. However, you may not necessarily realize or remember that you possess these attributes, either because you take them for granted or because you just weren't aware of them.

A good place to begin is the exercise for identifying strengths that we showed you in chapter 1. It will help you to become aware of your strengths and resources, or you can add to the list you have already been accumulating. Calling on your strengths and resources makes lasting change much more attainable.

Take Small Steps and Reward Yourself for Small Improvements along the Way

Patients often talk about how long and arduous the path to healing seems sometimes. Gradual improvements are very important, but sometimes it's hard to gauge how far you've actually come. Patty's observations capture an important aspect of this.

Patty Seeks a Sign

Patty was a chronic gut sufferer who dedicated herself to daily practice of the various techniques in the CORE toolkit. She carried out the relaxation exercises, Neurohormonal Retraining, and physical exercise. She watched her diet and tried to stay aware of her various stresses so she could do something about them. She knew she felt better, but she didn't know exactly how much she had improved.

"It feels hard to tell how my progress is going," Patty complained. "It's almost like I'm swimming in the ocean—I might be making progress and swimming a long way, but it's hard to tell how far I've come because there are no distance markers to show how much I've done. All I can see when I look for my bearings is more and more water on the horizon."

Patty's comments highlight an important point. Even though you are making progress in the larger picture, day-to-day changes may be imperceptible. Recognize and reward each forward step, no matter

Recognize and reward each forward step, no matter how small.

how small. Be sure to write down in a journal when you have maintained a certain practice for a week, and highlight it. These markers let you know that progress is proceeding.

For example, if you're starting breathing exercises, set your sights on small goals first. Perhaps aim for doing five minutes of a breathing technique (say, 4-4-8 breathing) three days a week. And then, when you've maintained that goal successfully for a week, give yourself some acknowledgment and also a small reward.

What kind of reward? Something that is pleasurable and enjoyable for you—and hopefully healthy. Go to hear a music performance you enjoy. Put a few dollars aside for a little trip to a bed and breakfast for the weekend. Buy some new flowers or plants for your garden. These positive reinforcements provide encouragement for keeping up the good work, and they serve as a reminder that you are indeed making progress.

Pay Very Close Attention to Your Improvements

When you go through a period of improvement, and you're feeling more improvement and relief from gut symptoms, it's important to acknowledge your relief. What feels different? Is the bloating a little less? Is there less gas pain? What does *that* feel like? What does it really feel like when you wake up in the morning and the abdominal pain is at a one out of ten instead of a six out of ten? This information is vital because many chronic gut sufferers go through long periods of time when their symptoms feel unchanged for what feels like months. This is why so many of our patients with IBS feel frustrated and hopeless by the time they get to us.

So when you're feeling better, when the cramping and bloating feel decreased, when the diarrhea or constipation has eased for a while, *pay attention to this*. This becomes another vitally important awareness exercise. Ask yourself the following questions when your symptoms are improved, and write the answers down in your CORE journal:

- When I'm feeling better, what is my emotional state?
- How are my thinking patterns different?
- How has my feeling of hopefulness about the future changed?
- How has my level of motivation for doing self-care activities and exercises changed?
- How do I view myself differently as a person?

The answers to these questions provide valuable reference points that you can store internally. Then if you have a setback later on, you can refer back to these guideposts of success. You can say to yourself, "I feel lousy at the moment with this sharp abdominal pain and indigestion. But I can also clearly remember what it's like to feel better, so I know I can get back there again."

Practice, Practice, Practice

Several years ago Dr. Weisberg was taking a class in Tai Chi, a martial art that focuses on physical grounding, balance, concentration, and a series of gradual, precise movements. In fact, a big part of the reason that he took the class is because he was having difficulty establishing balance in his life. As his practice got busier, his days were becoming fuller. He felt his daily schedule getting progressively out of control. He really enjoyed learning the steps of the Tai Chi form and found it energizing and relaxing when he would make it to our weekly sessions. Our instructor, Sensei Chen, reminded the students to be sure and practice the whole set of Tai Chi postures and movements for at least twenty minutes per day, every day.

Of course, Dr. Weisberg had every intention of following Sensei Chen's instructions for regular practice, and for the first few weeks, he followed through successfully. But as the weeks went on, it got harder. Sometimes his patients had emergencies he had to attend to late at night. Sometimes he had plans with friends that he had not seen for a long time, and he went to see them instead of practicing. Sometimes he was short on sleep after a number of long days, so he slept for an extra twenty minutes rather than practicing Tai Chi. Other times, he just felt so tense and preoccupied after a long week that when the weekend came, he was too wound up to do the practice. Instead, he just watched television or went to a movie.

When he returned to class the following week, it turned out that he wasn't the only student in class who struggled with practice. Most of his fellow students found it increasingly difficult to keep with the routine. As Sensei Chen heard their excuses for not practicing, he smiled gently and said this: "If you're tired, do a tired Tai Chi practice. If you're tense, do a tense Tai Chi practice. If you're busy, do a busy Tai Chi practice. If you're bored, do a bored Tai Chi practice. But whatever you're feeling, keep doing your Tai Chi practice."

The more consistent your practice is, the easier it will be for the new behaviors and skills to stick. For example, if you are going to begin using a food diary, commit to doing it at the same times, in the same place at your dining table, for at least two weeks. When cues like time of day, place, and circumstances are the same every time, it is easier for the new behavioral pattern and physiological change to become more permanent.

On some days, you may feel too busy, too tense, too preoccupied, too bored, or too tired to maintain regular practice of these new skills. But regular practice pays off, and with time you will teach your body/mind new habits that will become more permanent patterns of wellness. As Sensei Chen taught Dr. Weisberg, however you feel, keep doing your practice.

It will get easier with time, and the benefits for your healing and relief will grow as your practice becomes more of a part of your daily life.

Build Resilience

Resilience is the capacity to adapt well in the face of adversity, tragedy, threats, or even significant sources of stress—such as family and relationship problems, serious health issues, or workplace and financial worries. Resilience is the capacity to bounce back from these difficult experiences. This makes it easier to maintain long-term improvements in your lifestyle and in your digestive distress.

Being resilient in the face of gut distress does not mean that you don't go through episodes when your symptoms make you feel drained, depressed, or discouraged. Emotional pain and sadness are common in people who have suffered major adversity or chronic illness in their lives. In fact, the road to resilience sometimes involves considerable emotional distress. But resilience is a fairly common human attribute, and anyone can learn the behaviors, thoughts, and actions they need to bounce back.

Factors in Resilience

Your CORE program for healing already contributes to the development of resilience, but here are some additional variables that are associated with a greater ability to bounce back:

- The capacity to make realistic plans
- The ability to carry out your plans
- A positive view of yourself
- Confidence in your strengths and abilities
- Skills in communication
- Skills in problem solving
- The capacity to manage strong feelings and impulses

How You Can Build and Increase Your Resilience

Attention to the following skills and attitudes will help you build and increase your resilience:

- Increase self-awareness: Understanding, noticing, and accepting bodily reactions increase self-awareness and resilience. This is yet another benefit of practicing Neurohormonal Retraining, which increases your resilience over time. Resilient people are aware of their situation, their own emotional reactions, and the behavior of those around them. In order to manage feelings, it is essential to understand what is causing them. By remaining aware, resilient people can maintain their control of the situation and think of new ways to tackle problems.

- Accept occasional setbacks: Another characteristic of resilience is the understanding that life is full of challenges. While we cannot avoid many of these problems, we can remain open, flexible, and willing to adapt to change. When you learn to accept this as part of the healing process, you will develop better coping skills over time, which in turn reduces the hypersensitivity of the nervous system that aggravates gut symptoms.

- Develop control: Do you perceive yourself as having control over your own life? Or do you feel like much of what goes on in your life happens to you without you having much to say about it? Generally, resilient people tend to have what psychologists call an internal locus of control. They believe that the actions they take will affect the outcome of an event. Of course, some factors are simply outside of our personal control, such as natural disasters or the reactions of other people. While we may be able to put some blame on external causes, it is important to feel that we have the power to make choices that will affect our situation, our ability to cope, and our future.

- Develop problem-solving skills: Effective problem-solving skills are essential. When a crisis emerges, resilient people are able to formulate a solution that will lead to a safe outcome. In troubling or frustrating situations, people sometimes develop tunnel vision. They fail to note important details or take advantages of opportunities. Resilient individuals, on the other hand, are able to calmly and rationally look at a problem and envision a successful solution.

- Establish strong social connections: Whenever you're dealing with a problem, it is important to have people who can offer support. Talking about the challenges you are facing can be an excellent way to gain perspective, look for new solutions, or simply express your emotions. Friends, family members, coworkers, and online support groups can all be potential sources of social connectivity.

- Ask for help: While being resourceful is an important part of resilience, it is also essential to know when to ask for help. During difficult times, people can benefit from the help not only of friends and family but also of psychologists and other mental health professionals specially trained to deal with demanding situations.

How to Weather Occasional Relapses

Part of the key to successful change is anticipating and being prepared for sporadic flare episodes or relapses. With this knowledge, you can have a plan in place for getting back on track with your CORE healing regimen. What follows here are specific suggestions for navigating this sometimes tricky path, starting with the story of Inga.

Inga Has a Setback

Inga was feeling very hopeful because of the progress she was making on her individualized CORE program for healing. Her diarrhea, cramping, and indigestion had become much less frequent and less severe.

One night, she went with some friends to a party. She was so relaxed and feeling so much better that she allowed herself to indulge in some snacks and beverages that she had previously identified as symptom triggers. Ordinarily, she wouldn't have done this, but she figured it wouldn't hurt her since she was doing so much better. She ate some pizza and washed it down with a beer.

Within twenty-four hours, she began to feel a flare-up of diarrhea and painful cramps that lasted for four days. When she came to her next appointment, she felt angry, frustrated, and discouraged. "Sometimes these flare-ups are so demoralizing that I just feel like giving up on taking care of myself," she confessed. "I get afraid that all my effort is not going to make any difference."

Dr. Weisberg explained to her that setbacks are common as part of the path to lasting healing, and that this didn't mean that her progress wouldn't continue.

She went on to describe her confusion. Inga had been an athlete for many years, running track in high school and college. "I have always known how to keep pushing harder in my training, how to get my times faster in the 400 meter. But for the life of me, I've never learned how to deal with the one-step-back part of two steps forward, one step back, a process my coach always talked about."

This notion of two steps forward and one step back is very important to understand when you are striving to achieve permanent healing for your gut distress. Humans seem to be hardwired to change in this type of pattern. It's almost as if once we make a new change, our body/mind system needs a little time to incorporate the change, and then it reverts back a little bit to the old way of feeling before permanently moving forward.

Please be assured that the changes that you make are not just a matter of simply a different probiotic or a different exercise or a different thought about your symptoms. The steps you are taking are truly changing your physiology. They are changing the cascades of neurotransmitters and hormones that form the communication pathway between your brain and your gut. You are taking advantage of positive neuroplasticity (see chapter 9), meaning that you are in the process of literally creating new, healthier neural pathways in your brain and nervous system. This takes time, and it is often an up-and-down process, rather than a straight line to a perfect gut.

Steps for Dealing with Symptom Flare-Ups and Temporary Setbacks

1. Breathe: First, take a deep breath. Maybe even two or three.

2. Recognize: Keep in mind that flare-ups and setbacks are a normal part of the healing process. It is common for your

symptoms to get a little worse after a period of improvement, and then they'll get better again.

3. Accept: Once you recognize that you're having a flare episode, the next step is to *accept* it rather than *fight* it. Reactions of anger, panic, frustration, and futility are normal, but they can also agitate your sensitized nervous system, heightening the pain and bloating even more. Don't be afraid of the heightened feelings of pain, bloating, cramping, or other gut distress that occur with this flare-up. Don't let them scare you. This doesn't mean you've lost your progress. It's a temporary one step back.

4. Allow: Make room for whatever difficult emotions may surface. Allow the feeling, remembering, "This is (scary, sad, frustrating), but it's a normal part of the process, and it will pass." This activates the important capacity for self-soothing of difficult emotions. (Refer back to chapter 10 for useful reminders on how to do this.)

5. Get support: Let the people who are close to you know about your CORE healing program and its steps. Then, if a setback occurs, turn to them for support and encouragement. Let them remind you that you're not alone in this process of healing and that the flare-up episode will pass.

6. Stay the course: Get back to the fundamentals and self-care skills of your CORE program. Do Neurohormonal Retraining to sit with the heightened sensations and reduce the agitation. It's particularly helpful during flare-ups, too. Maintain your Ecological Rebalancing plan, sticking with the right foods, probiotics, and nutrients that are part of your individualized plan for healing. Reduce whatever stresses may be flared up in any of the five categories of stress. Make sure that you're allowing enough time for sleep. Use your new skills for taking the sting out of any particularly difficult emotions that might be aroused during this flare-up.

7. Use the setback to your advantage: Although a flare-up can be frustrating or troubling, it also allows for the opportunity to gather valuable new information to add to your individualized CORE profile. You may become aware that some additional foods or stresses may affect your digestion in ways that you didn't know before the flare-up. Also, when you practice your Neurohormonal Retraining during the flare-up, you may become aware of a useful clue about something that may have contributed to the flare-up that you can avoid in the future.

8. Focus on strengths: Remember your unique strengths that you identified earlier in the book. Use this awareness for reassurance that you will make it successfully through this temporary challenge, just as you have made it through earlier challenges and difficulties in your life.

9. Take heart: It will get better.

Inga Gets Resilient

After receiving some reassurance and reminders about the steps of getting through a flare-up episode successfully, Inga calmed down and regained some of her confidence. Dr. Weisberg worked with responding to the flare-up symptoms. Starting with Neurohormonal Retraining practice, he encouraged Inga to find the place in her lower abdomen where the cramping pain was most bothersome.

Like before, she imagined that she was going to pull up a chair and sit right next to the cramping sensation exactly the way it is, without trying to change it in any way. Eventually she was able to do this. After about three or four minutes, she said the cramping sensation had moved from her lower abdomen to her right side, and the intensity reduced from a seven to a three. She stayed with the sensation another ten minutes, just noticing it, not analyzing it. After time had passed, she was a little teary but felt relieved. Her next words were "I guess I don't have to be so angry and scared about this—it will pass."

Inga got back to some of the basics of her Ecological Rebalancing practices as well. She recognized that the alcohol from the beer and the tomato sauce and cheese from the pizza were trigger foods that aggravated her sensitive gut. This episode reminded her to stick more closely to her dietary regimen. She noted some other dietary triggers in the past few weeks that may have made her more vulnerable to a flare-up: milk products and sweetened carbohydrates, mostly from eating a lot of cake at a number of recent graduation parties.

At the next appointment, Inga reported, "I feel even better now. I don't like flare-up episodes any better than I used to, but I'm reassured now that they're a normal part of the healing process. I won't be quite as scared or discouraged the next time a setback occurs. I know I'm on the path to healing, even when I have one step back."

Establish and Maintain Your CORE Support Network

As we've been saying all along, the lessons from clinical experience and research in recovery from chronic illnesses couldn't be more clear. When you find and engage the right amount and types of social support, you recover more quickly and more thoroughly from all kinds of difficult conditions, and you maintain new patterns of self-care behaviors easier and more quickly for the long run. Whether it's from a spouse, friends, neighbors, a support group, health care professionals, a spiritual/religious community, or other healthy sources, be sure to engage lots of support for the long-lasting lifestyle changes that will keep you on a healing track for years to come.

Resistance to Change Is Normal: Allow for It

It is a natural tendency for people to be oblivious about much of their lives. Whether it's avoiding their emotions or ignoring their daily habits, it's quite normal to resist being aware of yourself.

To clarify this, let's start by defining awareness. Awareness is the state or ability to perceive, to feel, or to be conscious of events, objects, or sensory patterns in your body. Sensory awareness means that you are able to feel the different sensations in your body as well as

Emotional and sensory awareness are closely interrelated, as we can only experience emotions and feelings in our body.

the changes and fluctuations that occur. Emotional awareness means you are able to recognize and experience the various emotional states you go through as well as the changes. Emotional and sensory awareness are closely interrelated, as we can only experience emotions and feelings in our body.

Why Do We Resist Awareness and Change?

We all have resistance to change and to the awareness of here-and-now experience. It's normal, and that must be understood in order for you to make changes to improve your gut health.

Change is hard for many reasons:

- It's unknown: One of life's greatest fears is of the unknown. We resist things when we don't know what is going to happen, and that leads to more anxiety.

- It's challenging: Change stretches us out of our comfort zone. Some people tolerate being stretched more than other people do.

- It's uncertain: When we change, we often enter untested waters. We prefer certainty and can experience anxiety when that certainty and constancy is challenged.

- It's unpopular: The resistance to change is universal. Change invites animosity and resistance from other people.

- It requires effort: Many people avoid making changes that affect their health and lifestyle because those steps require them to alter habits and invest more time and attention in how they feel and react in daily life.

Ambivalence

We have all had the experience of wanting to do something and at the same time wanting *not* to do it. This is part of human nature, and we see it in our practices all the time. For example, right now you might feel like you need to get some yard work done. At the same time, you may feel like nothing is more important than stretching out in your recliner and enjoying some music. This internal tug-of-war is called ambivalence.

Ambivalence shows up all the time when people are trying to make changes in behaviors, feelings, and habits regarding how they take care of their health. We commonly see ambivalence in patients who are looking for relief from chronic gastric distress. On one hand, you want to do whatever is necessary to feel better. On the other hand, you may feel on some days like there just isn't enough time to do the stretching, breathing exercises, observing, and dietary adjustments that are necessary for change.

How to Deal with the Normal Resistance to Change and Awareness

When you're dealing with normal resistance, first recognize that resistance to change is normal. It must be expected and accepted, and you must notice and acknowledge your concerns about making lifestyle changes. Knowing that this is part of the process can be helpful because you won't feel discouraged when it's hard to keep on the track of developing awareness.

Here are some additional strategies to deal with normal resistance to change and developing self-awareness, so you can stay on track with your progress.

- Remember why you're making these changes: You have been through many months or even years of suffering with symptoms of diarrhea, constipation, cramping, bloating, indigestion, and other debilitating gut symptoms. That level of distress is your best antidote to inaction and staying stuck. Distress is ultimately the best motivator for change. When you feel that your symptoms have been there long enough, when you feel that you are finally fed up with the daily disruption caused by these symptoms, then you will feel less ambivalent and more motivated to take steps to relieve your gut distress. This is why centering is the first step of the CORE process. Remember all that you've been through while trying to cope with IBS. Visualize the healthier, happier you who is free of distress. This is why you're going through the journey of learning these new skills. Keep in mind that as you develop more awareness of your sensations and emotions, you'll benefit from being able to change your body's habitual response to pain, bloating, and cramping. This will allow you to benefit from your body's intelligent signals. You're learning to trust your gut.

- Take it slow: As you practice the CORE skills for developing awareness and change, take it slow. Start by focusing on a non-painful sensation for a short time—just a minute or two. Notice its qualities without trying to change it. At this point, you're just learning to observe—you're not working hard to change anything. Later, you can move on to noticing sensations that are mildly uncomfortable.

  Take it slow as you're learning to notice emotions as well. Remember, feelings tie in to the pain circuitry that affects your gut symptoms, and this is why you're learning to recognize them. It's not that your symptoms are psychosomatic, but rather the brain processes sensation and emotion together. Start by allowing a few minutes to sit with a journal and notice what emotion or feeling is present in your body.

  If you have trouble naming or identifying feelings, you can start by using the simple feelings acronym SASHET, which stands for

  - Sad
  - Angry

- Scared
- Happy
- Excited
- Tender

Sit for a short time, and see if you can recognize one of the feelings from this list and just mark it down in your journal. As you practice more, you'll probably be able to add several more emotions to the list.

- Give yourself some credit: When you are able to notice and start writing down sensations or emotions, give yourself some credit! This is a new skill for many people, and you're making progress. For many people, this is a learned skill rather than something that comes naturally, and it will get better with practice.

- Two steps forward, one step back: One thing that we know from many years of treating patients with digestive distress is that the process of change is not a straight line of continuous progress from start to finish. Rather, as we emphasized before, it's as if we're hardwired to make some progress, slip back a little, and then progress even further each time. Here in Minnesota, where the winters get pretty snowy, we have a saying for that: sometimes when your car tire is stuck in the snow, you have to rock backwards a little before you can move forward out of the rut. On some days, you'll feel more enthusiastic about your practice, and you'll be able to notice your sensations and emotions quite clearly. On other days, it may feel like you can't notice anything. That's all right—just keep at it. Expect and allow for these fluctuations.

- Take a day off. Part of what reduces resistance to change is making your change plan flexible. Every seven or eight days, take a day off from working on any of the CORE awareness exercises if you'd like. (Who knows—maybe you'll start to notice what it feels like on those days too!) Just be sure to stay away from pizza and beer if they're on your no-no list.

Review the CORE Program

The more you practice the CORE exercises, the more you will understand when you reread this book. For now, we will have a quick review of the four steps of the CORE program for recovery, reemphasizing the important new developments of Ecological Rebalancing and Neurohormonal Retraining. As you read and think about the CORE steps again now, you will start to understand them with a deeper sense of *knowing*. Our hope and intent here is to deepen your understanding of these concepts and to make the skills and tools even more useful, practical, and meaningful for you. All of this leads to you being able to more successfully trust your gut for healing.

Our goal in this book has been to help you achieve a deeper, more comprehensive, longer-lasting healing for your digestive distress. We knew that to accomplish this, we would have to go beyond the scope of most consumer books on gut distress, which primarily focus on dietary change and stress management. These components are necessary but insufficient for deep healing. There's more to it than that. And most importantly, our intention has been to bring you new healing technologies that emphasize your natural self-healing capacities, your most powerful and important resource of all.

As we first told you in the introduction, CORE is an acronym that stands for

Center

Observe

Restore

Ensure

These steps are progressive, meaning that each step is dependent on the knowledge, experience, and skills gained in the previous step.

Center

Centering can be thought of as a combination of methods used to calm and balance yourself physically, emotionally, and mentally. This harnesses and focuses your resources in preparation to perform an activity or undergo significant change. As we discussed, one goal of your digestive self-healing process is to become centered, especially since the gut is the center of your body/mind in terms of emotions, energy, and intuition. As you learn to trust your gut, your body/mind system automatically becomes more balanced. As this occurs, it becomes easier to pay attention and access your self-healing resources.

We also shared with you some specific strategies for becoming more centered:

- Quieting your body/mind
- Identifying your strengths and resources
- Calling on your inner wisdom and intuition to picture a future healthier you
- Skills for you to build your centering expertise right away

The capacity to center is the prerequisite for the next CORE step: observe.

Observe

One mantra repeated throughout the book is that you can't change it if you can't see it/feel it/sense it. The CORE method is strongly based on the internal self-healing resources that you possess. Self-awareness is a crucial skill for self-healing. This entails the capacity to *observe*. Observation is not just an intellectual process. It is an important skill that comes from a centered vantage point.

Part of this proficiency includes noticing aspects of your life and health functioning in much closer detail than you may have done before. Your observation and awareness of the nuances of your sleep, your stresses, your diet, and your health history are vital. This bio-information is necessary for preparing for the essential changes you will make in your gut functioning, your sense of balance, and your life.

The capacity to observe from a quiet vantage point is also required for one of the most important tools you gained from this book: Neurohormonal Retraining—the ability to sit and observe a painful gut sensation in a new and different way. This is one of the most powerful change techniques you can learn. It changes the alarm dysfunctions of your brain-gut communication system, allowing for your body/mind to change how it responds to painful but non-harmful signals.

Now that you have started developing the capacity for centering and observation, you can begin to benefit from the book's two new technologies: Neurohormonal Retraining and Ecological Rebalancing. These skills are integral for the capability to restore natural healing and balance to your digestion.

Restore

Self-healing means restoring your digestion to the healthy, comfortable, well-balanced system that was there before your symptoms ever developed. To restore means to apply a battery of new skills for reestablishing equilibrium to your body/mind. The chapters in the "Restore" section of the book represent the how-to skills of Ecological Rebalancing and Neurohormonal Retraining, the new technologies that bring lasting relief as more effective alternatives to pharmaceuticals that merely cover up the symptoms.

Ecological Rebalancing addresses the synergy, the vital interaction between your inner and outer ecologies. This includes tools for restoring your interior ecology, using strategies such as prebiotics, probiotics, and special diets. You also learned about effective tools for establishing improved balance in your exterior ecology: balancing your needs for self-care and relationship demands with work and with other important people in your life.

Neurohormonal Retraining is the powerful new technique to retrain the way your brain responds to the sensations in your gut so you can restore the normal channels of communication between your digestive system and your pain control systems in the brain and nervous system.

There are two other sets of skills we shared with you toward the goal of restoring a well-functioning gut. The first is based on the notion that you need two parallel but different kinds of strategies for taming gut distress. One component represents daily self-help tools for calming and balancing your mind, body, and nervous system. These skills efficiently help your body/mind optimize its capacities for healing as part of a self-care strategy. Practice some of these strategies every day to maintain healthy resilience, regardless of how you're feeling. The other component involves Neurohormonal Retraining. When you encounter more severe or troubling digestive distress, Neurohormonal Retraining is the first line of response for changing the disturbed, agitated communication between your troubled gut and your brain.

In the second set of restoring skills, we covered the sensitive but crucial topic of resolving difficult emotions and their physical effects. A flare-up of gut distress is often set off by an emotional episode such as grief, anger, anxiety, or frustration. This is a sensitive topic because many people with chronic gut distress go through lengthy, extensive medical workups only to be told that their symptoms are psychosomatic or just due to stress. Any mention of emotional factors can feel shameful or threatening. Difficult emotions do not *cause* IBS, but they can certainly affect brain-gut communication and stir up episodes of heightened cramping, bloating, gas, pain, and indigestion. The specific tools offered in this chapter can help you defuse the power of negative emotions so they don't disrupt your digestion—or your morale.

Ensure

The purpose of this last part of the CORE program is to ensure that you maintain your gains in self-care skills and healing your gut so that they become part of an ongoing healthy lifestyle rather than just a momentary change. You've learned a number of tools and tips to make sure the changes you make become long lasting and longstanding.

The Next Steps in Your Self-Healing Journey: Rebalancing and the Rhythms of Life

We have reached the end of this book, but it is merely the beginning of your journey for lifelong digestive healing. By learning and applying the skills, tools, and awareness gained in this book, you are now embarking on a process of dynamic self-care to ensure vibrant good health for now and for the future.

As Eastern medicine traditions have known for thousands of years, health comes from maintaining a dynamic balance. *Dynamic* means "movement and change, always shifting, progressing and balancing— not static or stagnant." Life keeps evolving; it doesn't sit still and wait for you. Life and health can best be understood as rhythms and cycles of balancing factors. Activity and rest. Work and play. Contraction and expansion. Day and night. Hot and cold. Stimulation and calm. This is the nature of the human body and the world. Self-healing is what happens when we observe, respond, and move with awareness of these natural rhythms.

Health comes from maintaining a dynamic balance.

Even the greatest performers in sports, academics, and the arts know that their progress does not stand still, even for one day. They cannot rest on their laurels. They have to keep practicing to keep their skills sharp. Basketball superstars continue to practice their three-point shots. Baseball legends continue to take batting practice. World-class violinists keep practicing their scales. The greatest ballet dancers in the world stretch and practice their craft every day. No one is ever done— it's a continual practice.

Your CORE program is not a linear process that starts at point A and ends at point B. Instead, it is a new rhythm in your life. It is a circle that you must continue to move with and carry out. You have to wake up every day and stretch out the stiffness of sleep. It doesn't matter that you stretched yesterday. Today is a new day and you begin with a new stretch. The same with diet. You may have followed a great diet yesterday, but you need to follow a great diet again today. You may have done a great job noticing and accepting the sensations of your unsettled abdomen yesterday. But today is a new day, and your body/mind is sending you intelligent daily signals that require your awareness and welcoming sense of observation again today.

You are never done centering yourself. You are never done observing and cultivating self-awareness. You are never done restoring your balance, and you are never done ensuring that you stay healthy. The skills in this book need to become habits, part of your new lifestyle for digestive health. The good news is that maintaining this new daily routine becomes more and more enjoyable as you embrace it.

The reward for engaging in this ongoing practice is the profound satisfaction and hopefulness that comes from unleashing your own self-healing resources—the most powerful medicine of all.

ACKNOWLEDGMENTS

Co-authoring a consumer book is a demanding, challenging, and sometimes a disorienting experience. A community of professionals cheered us on and provided help, skills, and a much-needed roadmap to our goal, and we want to express our mutual gratitude to them.

We are deeply appreciative of our professional colleagues who reviewed elements of our manuscript. Their generous donation of time and valuable feedback was crucial in informing and shaping the final product. They include Neville Basman, MD; Jennifer Blair, LAc, MAOM; Suzanne Candell, PhD; Alfred Clavel, MD; Robert Decker, LAc, RPh; Carolyn Denton, MA, LN; Henry Emmons, MD; Penny George, PsyD; Stephen Gilberstadt, MD; Bill Manahan, MD; Julie Stroud and Carolyn Torkelson, MD, MS.

Greg's Personal Acknowledgments

My contributions to this book have been possible because of many wonderful people in my life. We all need someone who believes in us, and, despite our shortcomings, sees potential and encourages us, who even makes deep sacrifices for us. Thank you Shawn. Thank you Penny. Thank you friends. Thank you from the bottom of my heart.

Today I am privileged to work with dozens of absolutely wonderful healers in Allina Health's Penny George Institute for Health and Healing. Each has touched and blessed this book in some way. Thank you. I am a much better physician today because of you. Much of the material for this book also derives from my six years on the faculty of the Keio University School of Medicine, Tokyo, Japan. I am deeply grateful to all of my teachers at Keio, the Japanese Society of Oriental Medicine (JSOM), and the United States-Japan Foundation Leadership Program. The kindness and generosity of my Japanese colleagues know no bounds. In particular, I want to highlight my mentor Kenji Watanabe, MD, PhD, FACP, Director of Keio's Center for Kampo Medicine. Critically important, I was able to read, write, and teach in Japanese because of my very talented tutor, Ms. Kumi Yoshiga, a true angel of patience.

I am also deeply grateful for the collaborative relationship between the Minneapolis College of Art and Design (MCAD) and Allina Health forged by their respective leaders President Jay Coogan and CEO Ken Paulus. I had the unique opportunity to work in the clinic of the Penny George Institute with five very creative MCAD students: Jenna Ballinger, Anthony Konigbagbe, John Kozak, Jenny Kunstel, and Megan Leitschuh. They helped bring forth elements of chapter 6. Their work was possible because of the program's organizers, Jess Roberts, M.Arch, Senior Design Strategist for Allina Health's Center for Healthcare Innovation, and MCAD Professors Jerry Allan, MA, and J. Kevin Byrne, MFA, MA. My continuing work with senior extern John Kozak resulted in additional contributions to this book.

Mark's Personal Acknowledgments

I want to thank some of my close friends and colleagues who provided support, encouragement, and inspiration through this amazing and sometimes arduous journey. These include Suzanne Candell, PhD; Carolyn Daitch, PhD; Louis Damis, PhD; Sheryll Daniel, PhD; Steven Gurgevich, PhD and Stephen Pannebianco, MD. To Alfred Clavel, MD: through our personal and professional collaboration over the last 20 years I have benefited tremendously, learned valuable lessons about the true nature of integrative medicine, and will always be grateful.

To Belleruth Naparstek, LISW, BCD: thank you for your encouragement when I first brought up my ideas for a book several years ago, and thanks also for suggesting I meet Caroline Pincus, your "book midwife," who eventually became our publisher.

To Marc Schoen, PhD: your friendship, guidance, and insights into the process of bringing a book to fruition have been invaluable. You have been my writing mentor, and I feel so fortunate to have gained from your support, expertise, and experience as a published author.

To Lauri: your love, encouragement, and reassurance helped me weather the tumultuous ups and downs of this creative process.

And to my daughter Emma: you have always been a shining light in my life. Your creativity and spirit remain a source of delight to me today and always.

Concluding Acknowledgments

We want to acknowledge Executive Editor and Associate Publisher Caroline Pincus as well as the rest of the great team at Conari Press/Red Wheel. Caroline has enthusiastically advocated for our project from the inception, and the rest of the team has done a great job of helping move the project along to completion.

Katharine Sands, our literary agent at the Sarah Jane Freymann Agency, saw the value in our collaboration from early on and provided steadfast support, cheerleading, and occasional prodding when needed. She intervened at crucial turning points in the development of the book, and we greatly appreciate her contributions.

Several ghostwriters have contributed to this effort. Thanks go out to Maryann Karinch and Barbara Barnett for their early input. Special thanks go to Gretchen Kelly, whose expertise, tireless efforts, and belief in our vision helped us produce a successful proposal, which led us to finding a wonderful publishing home for *Trust Your Gut*.

In particular, we are indebted to Steve LeBeau. Steve exhibited a unique gift for harnessing, catalyzing and combining the talents and skills of two very strong-minded and opinionated co-authors. He constantly challenged us to take our technical scientific and clinical concepts and break them down into simple, readable prose. Steve brought out the best in us, and this book would not have come to fruition without his skillful wordsmanship and diplomatic skills.

Finally, we wish to acknowledge that so many people have arrived at our clinics exasperated and hopeless. Their recoveries needed to be honored and shared. We wrote this book on evenings and weekends over several years to share the very important message that there is hope, that life will get better, that you can take control of your health. We wish to express our deep gratitude to our loved ones, family and friends, who patiently continued to provide support and encouragement during this difficult process. With a very deep bow, *Thank you.*

ABOUT THE AUTHORS

Gregory A. Plotnikoff, MD, MTS, FACP, is a board-certified internist and pediatrician who has received national and international honors for his work in cross-cultural and integrative medicine. He is a graduate of Carleton College, Harvard Divinity School, and the University of Minnesota Medical School. Plotnikoff is a frequent invited speaker and has been quoted in the *New York Times, Chicago Tribune, Los Angeles Times*, and numerous other newspapers internationally. He has been heard on Minnesota Public Radio's "All Things Considered," "Mid-Morning," and NPR's "Speaking of Faith" and "Science Friday." He has also been a frequent commentator on medical issues for local radio and television. He practices at the Penny George Institute for Health and Healing in Minneapolis, MN. Visit Dr. Plotnikoff at *www.gregory plotnikoff.com.*

Mark B. Weisberg, Ph.D., ABPP, is a clinical health psychologist. He is a Community Adjunct Professor in the Center for Spirituality and Healing, University of Minnesota, and is a Fellow of the American Psychological Association as well as a Fellow and former Vice President of the American Society of Clinical Hypnosis. He is a co-owner of the Minnesota Head and Neck Pain Clinic, an integrative chronic pain clinic with four locations in the Twin Cities. Additionally, he has a private practice in Minneapolis. Weisberg has been involved in clinical practice, teaching, and consultation in integrative mind-body medicine for the last twenty years. He has provided consultation to many integrative medicine programs around the country, including Dr. Andrew Weil's Program in Integrative Medicine at the University of Arizona Medical School. Weisberg is frequently interviewed on television, radio, print media, and the Internet. In addition to teaching online classes, he gives several invited keynote addresses every year and teaches at many national conferences. Visit him at *www.drmarkweisberg.com.*

TO OUR READERS

Conari Press, an imprint of Red Wheel/Weiser, publishes books on topics ranging from spirituality, personal growth, and relationships to women's issues, parenting, and social issues. Our mission is to publish quality books that will make a difference in people's lives—how we feel about ourselves and how we relate to one another. We value integrity, compassion, and receptivity, both in the books we publish and in the way we do business.

Our readers are our most important resource, and we appreciate your input, suggestions, and ideas about what you would like to see published.

Visit our website at *www.redwheelweiser.com* to learn about our upcoming books and free downloads, and be sure to go to *www.redwheelweiser.com/newsletter* to sign up for newsletters and exclusive offers.

You can also contact us at *info@redwheelweiser.com*.

Conari Press
an imprint of Red Wheel/Weiser, LLC
665 Third Street, Suite 400
San Francisco, CA 94107